THE DESIGN REVOLUTION

ANSWERING THE TOUGHEST

QUESTIONS ABOUT

INTELLIGENT

DESIGN

William A. Dembski

IVP Books

An imprint of InterVarsity Press
Downers Grove, Illinois

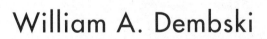

ivp

Inter-Varsity Press
Nottingham, England

InterVarsity Press
P.O. Box 1400, Downers Grove, IL 60515-1426 USA
World Wide Web: www.ivpress.com
E-mail: email@ivpress.com

Inter-Varsity Press, England
Norton Street, Nottingham NG7 3HR, England
Website: www.ivpbooks.com
E-mail: ivp@ivpbooks.com

InterVarsity Press® is the book-publishing division of InterVarsity Christian Fellowship/USA®, a student movement active on campus at hundreds of universities, colleges and schools of nursing in the United States of America, and a member movement of the International Fellowship of Evangelical Students. For information about local and regional activities, write Public Relations Dept., InterVarsity Christian Fellowship/USA, 6400 Schroeder Rd., P.O. Box 7895, Madison, WI 53707-7895, or visit the IVCF website at <www.intervarsity.org>.

Inter-Varsity Press, England, is the book-publishing division of the Universities and Colleges Christian Fellowship (formerly the Inter-Varsity Fellowship), a student movement linking Christian Unions in universities and colleges throughout the United Kingdom and the Republic of Ireland, and a member movement of the International Fellowship of Evangelical Students. For information about local and national activities write to UCCF, 38 De Montfort Street, Leicester LE1 7GP, email us at email@uccf.org.uk, or visit the UCCF website at www.uccf.org.uk.

All Scripture quotations, unless otherwise indicated, are taken from the Holy Bible, New International Version®. NIV®. Copyright ©1973, 1978, 1984 by International Bible Society. Used by permission of Zondervan Publishing House. Distributed in the U.K. by permission of Hodder and Stoughton Ltd. All rights reserved. "NIV" is a registered trademark of International Bible Society. UK trademark number 1448790.

Design: Cindy Kiple

Images: Nautilus:Rick Franklin/IVP
       Galaxy: Kevin Kelley/Getty Images

USA ISBN-10: 0-8308-3216-5
    ISBN-13: 978-0-8308-3216-3
UK ISBN-10: 1-84474-014-5
    ISBN-13: 978-1-84474-014-7

Printed in the United States of America ∞

Library of Congress Cataloging-in-Publication Data has been requested.

British Library Cataloguing in Publication Date
A catalogue record for this book is available from the British Library.

| P | 20 | 19 | 18 | 17 | 16 | 15 | 14 | 13 | 12 | 11 | 10 | 9 | 8 | 7 | 6 | 5 | 4 | 3 | 2 |
| Y | 22 | 21 | 20 | 19 | 18 | 17 | 16 | 15 | 14 | 13 | 12 | 11 | 10 | 09 | 08 | 07 | 06 |

"For the past decade or so 'intelligent design' has stirred a storm of controversy. Is it nothing more than gussied-up creationism, as its critics charge, or a new scientific paradigm, as its advocates maintain? This sprightly catechism, written by the movement's leading theoretician, offers believers and skeptics alike (and I count myself among the latter) an authoritative, if one-sided, introduction to what the fuss is all about."

RONALD L. NUMBERS, COLEMAN PROFESSOR OF THE HISTORY OF SCIENCE AND MEDICINE, UNIVERSITY OF WISCONSIN

"When I first heard William Dembski lay out his fundamental ideas on design, my immediate thought was a disjunction: we have here either a stunning conceptual breakthrough or a brilliantly illuminating mistake. Conceptual work looks easy when you see the conclusions; it is maddeningly difficult to absorb, refute or evaluate; the temptation to cut to the chase and cry victory is enduring. I was not in the least surprised by the intellectual, academic, theological and cultural storm that followed the public presentation of Dembski's proposals. He has burrowed deep into a warren of issues that will take decades to unravel and evaluate. In this volume he does not shirk the crucial questions that have to be addressed from his side of the house. His admirers should heed the complexity and nuances of his position; his critics need to do justice to the specificity and intellectual sensitivity of his claims; anyone interested in the interaction of theology, philosophy and natural science should listen in very carefully before making up their minds."

WILLIAM J. ABRAHAM, UNIVERSITY DISTINGUISHED TEACHING PROFESSOR, SOUTHERN METHODIST UNIVERSITY

"Bill Dembski has come a long way in five years from the very technical approach of his *Design Inference*. One of his key concepts in arguing from the intrinsic features of a system to a designer is its 'specified complexity.' Here, through luminous examples provided throughout this volume, he makes this concept more concrete, clearer and much more persuasive, and links it very helpfully to information theory. Together with a lucid presentation of his arguments for a designing agency for living organisms, he deals with many of the objections to intelligent design theory, especially those that claim it involves supernatural causes, violates the laws of nature and is non-scientific. He does this very readably, in short snappy chapters, which will make his analysis accessible to a much wider audience. Dembski has the rare ability to create profound new concepts, and he is making rapid strides from the bright but limited scientist into a mature interdisciplinary thinker. I believe it is only a matter of time before he becomes a very formidable mind. At times, stung by the fevered polemics of his critics, he tends, like Samuel Johnson, to 'argue for victory,' or even to claim victory. This, I am sure, will pass."

JOHN ROCHE, LINACRE COLLEGE, OXFORD

"Mainstream modern science, with its analytical methods and its 'objective' teachings, is the dominant force in modern culture. If science simply discovered and taught the truth about reality, who could object? But mainstream science does not simply 'discover the truth'; instead it relies in part on a set of unscientific, false philosophical presuppositions as the basis for many of its conclusions. Thus, crucial aspects of what modern science teaches us are simply shabby philosophy dressed up in a white lab coat.

"In this important new book, Dr. Bill Dembski continues his groundbreaking effort to show just how unscientific many modern scientists tend to be. If we are truly open to all the evidence, we can discover by the use of our unaided reason that the natural world is not the purposeless outcome of law—itself of unknown origin—and chance. This revolutionary approach has broad implications for science and broader implications for modern culture. Among many other things, Dr. Dembski's book is further evidence of the critical need for students in our public school systems to learn what is really going on in the disputes at the cutting edge of science rather than having their understanding of the natural world veiled and distorted by the prejudices of the past."

SENATOR RICK SANTORUM, UNITED STATES SENATE

"William Dembski is asking, and forcing the rest of us to confront, a profoundly important question: Is nature a closed system of efficient and material causes? Some people, purporting to defend the honor and integrity of science, immediately say: 'Of course, it is!' To them, Dembski issues a challenge: Does scientific inquiry itself vindicate the proposition that nature is a closed system of efficient and material causes? Or is this proposition accepted by many scientists and others as a matter of faith? Some who acknowledge that 'scientific naturalism' is a set of philosophical assumptions, rather than a fact that can be demonstrated by scientific methods, defend it on the grounds that the success of scientific inquiry based on it provides ample ground for its rational affirmation. From them, Dembski demands proof that scientific inquiry and the knowledge it generates actually do presuppose the exclusion of causes beyond the efficient and material. This proof has, it must be said, not been forthcoming.

"In his boldest move, Dembski argues that careful attention to the various manifestations of order discovered in the natural world suggests that efficient and material causes are in fact insufficient to explain the data into which the sciences inquire. We can reasonably infer from the order on display in nature that intelligence has figured in the design of natural phenomena and the natural world as a whole. It will not do for those to whom Dembski has issued his challenge to rely on their standing or authority within the scientific and academic establishments to wave him away. The truth is that the honor and integrity of science really are at stake in this matter. They would be profoundly tarnished by a dogmatic refusal to face up to Dembski's questions."

ROBERT P. GEORGE, MCCORMICK PROFESSOR OF JURISPRUDENCE
AND DIRECTOR OF THE JAMES MADISON PROGRAM IN AMERICAN IDEALS AND INSTITUTIONS, PRINCETON UNIVERSITY

"I find William Dembski's writing and argumentation on behalf of intelligent design to be careful, erudite, thorough and a formidable challenge to the theistic evolution camp I normally defend."

TED PETERS, PROFESSOR OF SYSTEMATIC THEOLOGY, PACIFIC LUTHERAN THEOLOGICAL SEMINARY AND THE GRADUATE THEOLOGICAL UNION

"The Design Revolution is about questions of fundamental importance: Can one formulate objective criteria for recognizing design? What do such criteria tell us about design in the biological realm? Sad to say, even to raise such questions is dangerous; but fortunately Dembski is not deterred. In this courageous book he takes aim at the intellectual complacency that too often smothers serious and unprejudiced discussion of these questions."

STEPHEN BARR, PROFESSOR OF PHYSICS, UNIVERSITY OF DELAWARE, AUTHOR OF *MODERN PHYSICS AND ANCIENT FAITH*

"Dembski's latest book indicates more clearly than any other recent publication that I know why CSI—the sort of order observed in complex machines or computer programs—cannot originate by cumulative selection. The improbability is far too great! Obviously, if biological systems contain considerable amounts of CSI as Dembski claims, then the standard Darwinian explanation is deeply flawed and what is needed is a new paradigm for understanding the natural world."

MICHAEL DENTON, MOLECULAR GENETICIST, AUTHOR OF *EVOLUTION: A THEORY IN CRISIS* AND *NATURE'S DESTINY*

"Not everyone will agree with all that is claimed in this book, and some knuckleheads will reject its theses without reading the book at all; but something of what is claimed in these pages will strike every fair-minded reader as important and provocative. The book does what it proposes to do. It meets criticisms of intelligent design honorably; it allows new ideas to breathe."

DAVID BERLINSKI, MATHEMATICIAN, AUTHOR OF *A TOUR OF THE CALCULUS* AND *SECRETS OF THE VAULTED SKY*

"This lucid and authoritative book attempts to answer the many objections that have been raised against the theory of intelligent design. Dembski argues powerfully that it is an intellectually defensible science. It should be read by anyone interested in the character of the world we live in, whether they wish to attack or defend the theory."

ROGER TRIGG, PROFESSOR OF PHILOSOPHY, UNIVERSITY OF WARWICK

"Through his techniques of design detection as well as his organizational efforts, William Dembski is making the revolution he describes in this book happen. If Thomas Huxley was 'Darwin's Bulldog,' Dembski is the man with the leash and the obedience training techniques to bring Darwinism into check."

EDWARD SISSON, LAW FIRM PARTNER, WASHINGTON, D.C.

"In this book, Dembski marshals the evidence for intelligent design, clarifying its detection as well as its potential impact on science. This book comes at an opportune time as the debate intensifies over naturalism's role in fully explaining (or not) intelligence and very complex information-bearing systems. The author's commitment to open the natural sciences to intelligent causation will generate lively discussion for a readership spanning many disciplines."

CONRAD JOHANSON, PROFESSOR OF CLINICAL NEUROSCIENCE,
BROWN UNIVERSITY

"The view that intelligence must be intrinsic to the essential nature of the universe is a perspective that is slowly gaining a wider acceptance among serious scholars within the community of scientists and philosophers. For such an inference concerning the intrinsic effects of intelligence on nature and her laws to be truly scientifically acceptable, however, requires that it be placed within a rigorous theoretical framework and that specific criteria for validating the truth of it be clearly enunciated. It is precisely this task that William Dembski undertakes in *The Design Revolution*. He asks the hard questions and answers them in a clear, frank and straightforward way. His responses to the toughest of queries concerning the investigation of intelligent phenomena intrinsic to the natural world are both lucid and sophisticated. He meaningfully moves the community of scientists and scholars along the path of investigating and understanding the momentous possibility that intelligence is an irreducible aspect of reality."

JEFFREY M. SCHWARTZ, M.D., RESEARCH PROFESSOR,
UCLA DEPARTMENT OF PSYCHIATRY, AUTHOR OF *THE MIND AND THE BRAIN*

To John and Dorothy Van Gorp,

my wife's parents,

salt of the earth

People almost invariably arrive at their beliefs

not on the basis of proof but on the basis of

what they find attractive.

BLAISE PASCAL, *THE ART OF PERSUASION*

CONTENTS

PART TWO: DETECTING DESIGN

PART THREE: INFORMATION

PART FOUR: ISSUES ARISING FROM NATURALISM

PART FIVE: THEORETICAL CHALLENGES
TO INTELLIGENT DESIGN

PART SIX: A NEW KIND OF SCIENCE

FOREWORD

Charles W. Colson

BILL DEMBSKI IS, ABOVE ALL, A REVOLUTIONARY. And this is a revolutionary book.

For years—far too many years—Darwinian evolution, the prevailing orthodoxy in the academy, faced no meaningful challenges. Those who believed in any other theory of biological origins were dismissed as religious cranks or fools. This is now beginning to change.

Bill Dembski has been in the vanguard of an exciting movement of thinkers, Christian and non-Christian, who effectively argue that naturalistic evolution can give no answers to the most vital questions of the day. In this book, Dembski delivers a stunning rebuttal of the idea that we live in a chance-driven, naturalistic universe and that time plus chance plus matter entails life in all its glorious complexity.

Immanuel Kant provides a convenient lens for understanding the current quandary. Kant was a theist deeply influenced by Christian pietism. As a philosopher, he made a radical proposal for epistemology, the branch of philosophy that studies how we know what we know. The upshot of his proposal was that there were two kinds of knowledge: that which could be determined as fact, that is, phenomenological knowledge; and that which could only be known by faith, that is, noumenological knowledge. This fact-faith distinction stuck and changed the way the Western mind approached the question of what it could know and not know.

Prior to Kant, people were perfectly willing to accept that since God created the universe, all truth was his and all truth could be known. We could rely upon God's authority and wisdom. As the pressures of the Enlightenment built, however, people surrendered the notion that God was necessary to explain creation. And having capitulated on this point, they readily surrendered the notion that God was necessary for the formulation of moral law or behavior. Over the years the fact-faith distinction became more firmly rooted so that, in the end, Western intellectuals insisted on basing both our science and our morality on naturalism.

At the same time, religious believers, bitten by the same Enlightenment bug, became increasingly private in their faith. Focusing on individualistic piety, believers forgot the holistic worldview thinking of previous generations. In adopting the fact-faith distinction, they compartmentalized their faith and cut it off from the rest of their understanding of the world. The result has been a wholesale abandonment of meaningful cultural engagement.

Such "two-story" thinking became almost unassailable and left the field wide open for naturalistic scientists to dominate Western thought—scientists who gave a naturalistic explanation of the biophyiscal universe with no reference at all to a creator or designer. There was religion on one hand and science on the other. And these two did not meet.

Although this was—and is—a false dichotomy, it has continued to dominate Western thinking even after naturalistic explanations for the creation of life began to fail. Today in public schools across America the idea that science provides a fully naturalistic explanation of the world and that faith is merely a matter of religion (or worse, a matter of "values"), which must be kept out of the classroom, is absolutely entrenched.

The intelligent design movement, of which Dembski is a key part, is effectively challenging this whole way of thinking. It has assaulted naturalistic evolution with lucid arguments and clear evidences of design.

The more we learn about the world in which we live, the more impressed we should become at what has been called the "anthropic principle." As I have written elsewhere, the anthropic principle states that in our own universe, all these seemingly arbitrary and unrelated features of the physical world—the distance of the earth from the sun, the physical properties of the earth, the structure of an atom—have one thing in common: they are precisely what is needed so that the world can sustain life. The entire biophysical universe appears to have been thought out and designed—intelligently designed.

Many scientists still hold on to the old two-story way of thinking and would rather not consider a thoughtful designer. Instead, they prefer to hold on to the naturalism that asserts a self-generating and self-explaining universe in which everything proceeds by chance and necessity, including the emergence of human life.

Dembski and such thinkers as Phillip Johnson, Michael Behe and Jonathan Wells have forced scientists to take seriously design and a designer. Their case is not based on the Bible or on religion. Instead, the case is based on scientific evidence. In place of naturalistic evolution, they are proposing a well-developed theory of intelligent design. Because it is a case of scientific theory versus scientific theory, secular thinkers are no longer able to simply dismiss design as a religious idea.

Dembski is a pioneer and a brilliant thinker who is making a tremendous mark. Not only are his ideas shaking intellectual circles, but they are now also filtering down to the popular consciousness. As a result, he is part of a movement to recapture the mind of our culture and to get intellectual balance back into the schools. This is one of the best and most hopeful things to come along in the Christian world in generations.

In *The Design Revolution*, Dembski covers a great deal of ground, answering objection after objection to intelligent design. In his years of writing, lecturing and debating intelligent design, he has heard just about every objection possible. In this book he takes these objections on one at a time, responding to the confused, the skeptical and the hostile. His arguments not only build the confidence of those of us who are already convinced of intelligent design but should also serve as a catalyst for serious thought by thoughtful skeptics.

Albert Einstein said, "I, at any rate, am convinced that God is not playing at dice." Indeed he is not. God carefully created a world that he cares for providentially. Dembski has, in this book, made that truth ever more clear.

PREFACE

EVER SINCE THOMAS KUHN PUBLISHED *The Structure of Scientific Revolutions* in the 1960s, just about every new idea in science has been touted as the latest scientific revolution. It's therefore not surprising that most scientific revolutions are overblown. I was part of one such overblown revolution in the late 1980s as a graduate student in Leo Kadanoff's physics lab at the University of Chicago. Chaos theory, also called nonlinear dynamics, was going to revolutionize science. A decade later, the promise and hype were largely spent. Yes, chaos theory offered some interesting insights into the interdependence and sensitivity to perturbation of physical processes. But after the revolution ran out of steam, our scientific conception of the world remained largely unchanged. Thanks to that experience, I take all declarations about the next big revolution in science with a stiff shot of skepticism.

Despite this, I grow progressively more convinced that intelligent design will revolutionize science and our conception of the world. To be sure, as a leading proponent of intelligent design, I have a certain stake in this matter. Nonetheless, there is good reason to think that intelligent design fits the bill as a full-scale scientific revolution. Indeed, not only is it challenging the grand idol of evolutionary biology (Darwinism), but it is also changing the ground rules by which the natural sciences are conducted. Ever since Darwin, the natural sciences have resisted the idea that intelligent causes could play a substantive, empirically significant role in the natural world. Intelligent causes might emerge out of a blind evolutionary process, but they were in no way fundamental to the operation of the world. Intelligent design challenges this exclusion of design from the natural sciences. In so doing, it promises to remake science and the world.

Revolutions are messy affairs. They are also far from inevitable. For there to be a revolution, there must be revolutionaries willing to put their necks on the line. They must be willing to take the abuse, ridicule and intimidation that the ruling elite can and will inflict. The ruling elite in this case are the dogmatic Darwinists and scientific naturalists. Rigidly committed to keeping intelligent causation outside the natural sciences, they misrepresent intelligent design at every step, charging that its critique of Darwinism (and of naturalistic theories of evolution more generally) is utterly misguided and groundless. Accordingly, the public is informed that intelligent design is religion masquerading as science or "Creationism in a Cheap Tuxedo" (the title of a newspaper headline). Moreover, the public is warned that intelligent design spells the death of science and that to teach intelligent design is intellectually (if not morally) in the same boat as teaching that the Holocaust didn't happen.

The acceptance of radical ideas that challenge the status quo (and Darwinism is as status quo as it gets) typically runs through several stages. According to Arthur Schopenhauer, "All truth passes through three stages. First, it is ridiculed. Second, it is violently opposed. Third, it is accepted as being self-evident." Similarly, evolutionist J. B. S. Haldane remarked, "Theories pass through four stages of acceptance: (*i*) this is worthless nonsense; (*ii*) this is an interesting, but perverse, point of view; (*iii*) this is true, but quite unimportant; (*iv*) I always said so."

I like to flesh out Haldane's four stages as follows. First, the idea is regarded as *preposterous:* the ruling elite feel little threat and, as much as possible, ignore the challenge, but when pressed they confidently assert that the idea is so absurd as not to merit consideration. Second, it is regarded as *pernicious:* the ruling elite can no longer ignore the challenge and must take active measures to suppress it, now loudly proclaiming that the idea is confused, irrational, reprehensible and even dangerous (thus adding a moral dimension to the debate). Third, it is regarded as *possible:* the ruling elite reluctantly admits that the idea is not entirely absurd but claims that at best it is of marginal interest; meanwhile, the mainstream realizes that the idea has far-reaching consequences and is far more important than previously recognized. And fourth, it is regarded as *plausible:* a new status quo has emerged, with the ruling elite taking credit for the idea and the mainstream unable to imagine how people in times past could have thought otherwise. With intelligent design, we are now at the transition from stage two to stage three—from pernicious to possible. This is the hardest transition.

The aim of this book is to facilitate the transition from stage two to stage three by giving supporters of intelligent design the tools they need to counter the attacks by critics of intelligent design. It is also intended for all honest skeptics of would-be scientific revolutions, for this book honors that healthy skepticism by fully and systematically responding to the toughest questions critics have raised concerning intelligent design. Readers will not need to grope about to find the questions or the answers. Nor will readers find tough questions missing in action.

In the past ten years, I've spoken at numerous colleges and universities on intelligent design, both in America and around the globe. I'm also regularly interviewed by the media about intelligent design. I have fielded an enormous variety of questions in both types of venues, and my work has drawn intense and extensive published criticism from the guardians of scientific orthodoxy. This book brings all those experiences—all those questions and their answers—together in one place. Think of this book as a handbook for replacing an outdated scientific paradigm (Darwinism) and as giving a new scientific paradigm (intelligent design) room to breathe, develop and prosper.

In speaking on intelligent design, I receive three types of questions. Often a question simply asks for further clarification. Sometimes, however, a question indicates a stumbling block that needs to be removed before further insight is possible. And finally, there is the question that is really not a question but rather an objection designed to "deep-six" intelligent design. I'll address all three types of questions in this book, but I'm particularly interested in the stumbling blocks. Intelligent design raises many stumbling blocks, especially for scientists and theologians. As much as possible, I want this book to remove those stumbling blocks. Clearing them away is presently the most important task in moving the design revolution forward.

Simply put, intelligent design is the science that studies signs of intelligence. Stated this way, intelligent design seems straightforward and unproblematic. Yet depending on where the intelligence makes itself evident, one may encounter fierce resistance to intelligent design. Archeologists attributing intelligent design to arrowheads or burial mounds is not controversial. But biologists attributing intelligent design to biological structures raises tremendous anxiety, not only in the scientific community but in the broader culture as well. Why is that?

C. S. Lewis, in his book *Miracles*, correctly placed the blame on naturalism. According to Lewis, naturalism is a toxin that pervades the air we breathe and an infection that has worked its way into our bones. Naturalism is the view that the physical world is a self-contained system that works by blind, unbroken natural laws. Naturalism doesn't come right out and say there's nothing beyond nature. Rather, it says that nothing beyond nature could have any conceivable relevance to what happens in nature. Naturalism's answer to theism is not atheism but benign neglect. People are welcome to believe in God, though not a God who makes a difference in the natural order.

Theism (whether Christian, Jewish or Muslim) holds that God by wisdom created the world. The origin of the world and its subsequent ordering thus result from the designing activity of an intelligent agent—God. Naturalism, on the other hand, allows no place for intelligent agency except at the end of a blind, purposeless material process. Within naturalism, any intelligence is an evolved intelligence. Moreover, the evolutionary process by which any such intelligence developed is itself blind and purposeless. As a consequence, naturalism makes intelligence not a basic creative force within nature but an evolutionary byproduct. In particular, humans (the natural objects best known to exhibit intelligence) are not the crown of creation, not the carefully designed outcome of a purposeful creator and certainly not creatures made in the image of a benevolent God. Rather, humans are an accident of natural history.

Naturalism is clearly a temptation for science, and indeed many scientists have succumbed to that temptation. The temptation of naturalism is a neat and tidy world in which everything is completely understandable in terms of well-defined rules or mechanisms characterized by natural laws. As a consequence, naturalism holds out the hope that science will provide a "theory of everything." Certainly this hope remains unfulfilled. The scandal of intelligent design is that it goes further, contending that this hope is unfulfillable. It therefore offends the hubris of naturalism. It says that intelligence is a fundamental aspect to the world and that any attempt to reduce intelligence to natural mechanisms cannot succeed. Naturalism wants nature to be an open book. But intelligences are not open books; they are writers of books, creators of novel information. They are free agents, and they can violate our fondest expectations.

There is an irony here. The naturalist's world, in which intelligence is not fundamental and the world is not designed, is supposedly a rational world because it proceeds by unbroken natural law: that is, cause precedes effect with inviolable regularity. On the other hand, the design theorist's world, in which intelligence is fundamental and the world is designed, is supposedly not a rational world because intelligence can do things that are unexpected. To allow an unevolved intelligence a place in the world is, according to naturalism, to send the world into a tailspin. It is to exchange unbroken natural law for caprice and thereby to destroy science. Thus, for the naturalist, the world is intelligible only if it starts off without intelligence and then evolves intelligence. If it starts out with intelligence and evolves intelligence because of a prior intelligence, then somehow the world becomes unintelligible.

The absurdity here is palpable. Only by means of our intelligence are science and our understanding of the world even possible. And yet the naturalist clings to this argument as a last and dying friend. This was brought home to me when I recently lectured at the University of Toronto. One biologist in the audience insisted I must take seriously that the world is two minutes old so long as I accept intelligent design. Presumably any creating intelligence could just as well create a deceptive world that appears old but was freshly created two minutes ago as create a verisimilitudinous world that appears old because it actually is old. That is certainly a logical possibility, but do we have any reason to believe it? Hundreds of years of successful scientific inquiry confirm a world that's structured to honestly yield up its secrets. If, further, the world reveals evidence of design, why should the mere possibility of a deceptive or capricious designer neutralize that evidence or lead us to disbelieve in the existence of a designer?

If we're going to take seriously the possibility of a designer misleading us, then we also need to take seriously the possibility of a natural world devoid of design misleading us. Imagine a natural world, devoid of design, where the laws of nature change radically from time to time, where time can back up and restart history on a different course, and where massive quantum fluctuations on a cosmic scale bring about galaxies that seem ancient but are in fact recent. It's not just designers that can be deceptive and capricious. The same is true of nature. Yet if science is to be possible, we need, as a regulative principle, to assume that nature is honest and dependable. And if nature is the product of design, that means we need, again as a regulative principle, to assume that the designer made nature to be honest and dependable.

It follows that the "two-minute-old universe" argument against intelligent design is an exercise in irrelevance. It cuts as much against naturalism as against intelligent design. And it can't even touch the point at issue, namely, whether certain biological systems are designed. To decide that question we must consult not theology or anti-theology but the evidence of biology. If that evidence points us to design, then that's where we must go. It would be absurd to say that the evidence points us to design but that we must nonetheless reject design because a deceptive designer might have designed the evidence to mislead us. That would be rejecting design by presupposing design.

When I pointed out to the Toronto biologist that Isaac Newton believed in intelligent design and didn't hold to a two-minute-old universe, he instantly remarked that Newton didn't know about evolution. Poor Sir Isaac. Presumably Darwin would have made him an intellectually fulfilled atheist and erased any vestige of intelligent design from his science (intelligent design figures substantively in Newton's *Principia*—see, for instance, his General Scholium). Somehow science and our knowledge of the natural world are supposed to unravel once we allow that intelligence could be a fundamental principle operating in the universe.

The charge that intelligent design spells the end of science and rationality is without merit. If anything, the very comprehensibility of the world points to an intelligence behind the world. Indeed, science would be impossible if our intelligence were not adapted to the intelligibility of the world. The match between our intelligence and the intelligibility of the world is no accident. Nor can it properly be attributed to natural selection, which places a premium on survival and reproduction and has no stake in truth or conscious thought. Indeed, meat-puppet robots are just fine as the output of a Darwinian evolutionary process.

I remarked that scientists wedded to naturalism have a hard time accepting intelligent design. Surprisingly, theologians often have an even harder time accepting intelligent design. Mainstream theology accepts the prevailing view that naturalism is a proper regulative principle for science—that science, to be science, must treat nature as a closed system of natural causes. Even if they are not metaphysical naturalists, mainstream theologians tend to be, therefore, methodological naturalists.

If this were their only reason for refusing intelligent design, then one would expect these theologians to hold methodological naturalism without ardor, as a mere working hypothesis. In fact, the idea that God could act not simply as some all-enveloping mushy influence but as an agent who makes a difference in space and time and takes responsibility for features of the world strikes many theologians as anathema. Often what's behind this distaste is an overdeveloped sensitivity to the evils of the world and a resulting compulsion to find an airtight theodicy. Theodicy attempts to justify the ways of God in the face of the world's evils. The easiest way to do this is not to let God get his hands dirty with the world. As a consequence, many theologians have a doubly hard time with intelligent design. Not only have they made their peace with a naturalistic construal of science, but they also have a theological need not to let divine action become too obvious or personal (e.g., if God acts here to do good, why doesn't he act there to prevent evil?).

This is not the book in which I address the theodicy problem (I plan to address it in a future book on Genesis, theodicy and the Christian doctrine of creation). Although theodicy is, to be sure, the thorniest problem facing theologians trying to make sense of intelligent design, it is not a problem for intelligent design per se. Intelligent design attempts to understand the evidence for intelligence in the natural world. The nature and, in particular, the moral characteristics of that intelligence constitute a separate inquiry. Intelligent design has theological implications, but it is not a theological enterprise. Theology does not own intelligent design. Intelligent design is not an evangelical Christian thing, or a generically Christian thing or even a generically theistic thing. Anyone willing to set aside naturalistic prejudices and consider the possibility of evidence for intelligence in the natural world is a friend of intelligent design. In my experience such friends have included Buddhists, Hindus, New Age thinkers, Jungians, parapsychologists, vitalists, Platonists and honest agnostics, to name but a few. As a consequence, intelligent design's fate does not stand or fall with whether one can furnish a satisfying theodicy.

Even though I'll be bracketing the theodicy problem throughout this book, I will nonetheless address certain criticisms of intelligent design motivated by it. According to design critic Edward Oakes, intelligent design makes the task of theodicy impossible. Why is that? Because, he claims, intelligent design is wedded to a crude interventionist conception of divine action and to a mechanistic metaphysics of nature. Neither of these criticisms is accurate. Intelligent design is compatible with just about any form of teleological guidance. Its concern is not with how a designing intelligence acts but with whether its action is discernible. Intelligent design therefore does not require an interventionist conception of design. As for intelligent design requiring a mechanistic metaphysics of nature, within the context of theology this is just the flip side of an interventionist metaphysics of divine action. Indeed, for God to be an intervening meddler requires a world that finds divine intervention meddlesome. Intelligent design requires neither a meddling God nor a meddled world. For that matter, it doesn't even require that there be a God. I address Oakes's concerns in chapter twenty ("Nature's Receptivity to Information") and chapter twenty-three ("Interventionism").

According to Oakes, the task of a Christian theodicy is to "show that an omnipotent and benevolent God can coexist with evil in His finite creation" (*First Things*, letter to the editor, April 2001). The key to resolving the theodicy problem for Oakes is Augustine's insight that God would not allow evil to exist unless God could bring good out of evil. Nevertheless, to speak of God bringing good out of evil could just be a fancy way of saying the end justifies the means. To avoid this charge, Oakes requires that the world be viewed "both as a totality and under the aegis of eschatology." In other words, God bringing good out of evil must be judged not on the basis of isolated happenings but on the basis of the totality of happenings as they relate to God's ultimate purposes for the world. All of this is sound Christian theodicy as far as it goes. I challenge Oakes and fellow critics to show that intelligent design, as developed in this book, conflicts with such a theodicy.

The theodicy question aside, how God relates to the theory of intelligent design requires one further clarification. Creationists and naturalists alike worry that when design theorists refer to a "designer" or "designing intelligence," and thus avoid explicitly referring to God, they are merely engaged in a rhetorical ploy. Accordingly, design theorists are saying what needs to be said to get skeptics to listen to their case. But as soon as skeptics buy their arguments for design, design theorists perform a bait-and-switch, identifying the designer with the God of religious faith. Whereas creationism is direct and forthright in its acknowledgment of God, intelligent design is thus said to be deceptive and sneaky.

This charge is unfounded. If design theorists are reticent about using the G-word, it has nothing to do with waiting for a more opportune time to slip it in. Insofar as design theorists do not bring up God, it is because design-theoretic reasoning does not warrant bringing up God. Design-theoretic reasoning tells us that certain patterns exhibited in nature reliably point us to a designing intelligence. But there's no inferential chain that leads from such finite design-conducing patterns in nature to the infinite personal transcendent creator God of the world's major theistic faiths. Who is the designer? As a Christian I hold that the Christian God is the ultimate source of design behind the universe (though this leaves open that God works through secondary causes, including derived intelligences). But there's no way for design inferences from physics or biology to reach that conclusion. Such inferences are compatible with Christian belief but do not entail it. Far from being coy or deceitful, when design theorists do not bring up God, it is because they are staying within the proper scope of their theory. Intelligent design is not creationism and it is not naturalism. Nor is it a compromise or synthesis of these positions. It simply follows the empirical evidence of design wherever it leads. Intelligent design is a third way.

When InterVarsity Press offered me a contract to write a sequel to my previous book, *Intelligent Design: The Bridge Between Science and Theology*, I was happy to sign it. *Intelligent Design* had been well received through InterVarsity, and so its editors urged me to write a sequel dealing with the most pressing issues confronting intelligent design. The most pressing issue at this time is to show that intelligent design is intellectually defensible and specifically that the criticisms and questions raised against it are answerable. Think of this book, therefore, as an extended question-and-answer period that helps clear the path for the design revolution.

Each chapter of this book opens with a question and is followed by an answer. I've tried as much as possible to make the chapters self-contained. This has necessitated some repetition, but I've kept it to a minimum. Although the questions in this book can be taken up separately, I have placed them in a logical progression so that the book can be read coherently from start to finish. I attempt to answer questions as I would in an audience setting—that is, in my own words, in plain English, and thus without extensive supporting quotes or technical apparatus. (The only notes and references occur in the text itself.) To be sure, writing my answers out allows me to be more thorough than I would be in a conversational setting. Nevertheless, I have attempted to keep my answers reasonably short. Chapters of many books tend to be around six to eight thousand words. Most of the answers in this book are around two thousand words.

Often when I write or speak about intelligent design and then step back to reflect on the fierce resistance my work receives, I'm reminded of those Kafka stories in which some hapless figure is tied up and smothered in endless bureaucratic red tape. The fundamental claim of intelligent design is straightforward and easily intelligible: namely, *there are natural systems that cannot be adequately explained in terms of undirected natural forces and that exhibit features which in any other circumstance we would attribute to intelligence.* That claim can be considered on its own merits. Let's look at some actual systems and do the analysis. This book is my attempt to cut through the red tape, psychological inertia and mental cobwebs that prevent intelligent design from receiving fair consideration. In short, it is my attempt at some much needed house cleaning.

Even so, my hopes for this book would fall short if a clean house were its only outcome. Besides cleaning house, this book aspires to provide a powerful new vision of science and the world, one that people will want to pursue because they find it so attractive. At the end of his *Origin of Species*, Darwin remarked that a person armed with his theory need "no longer look at an organic being as a savage looks at a ship, as at something wholly beyond his comprehension." At the time, Darwin offered a powerful vision for understanding biology and therewith the world. That vision is now faltering, and a new vision is offering to replace it. The new vision teaches us to see organic being as a civilized person would see a ship, namely, as the product of intelligent design. Nevertheless, we are to see its design not just intuitively; rather, we are to see it objectively, systematically and scientifically, as an engineer or architect who actually designed the ship. My hope is that this book will make such a new vision compelling.

For ideas to prosper, they must satisfy. In his *Art of Persuasion*, Blaise Pascal wrote, "People almost invariably arrive at their beliefs not on the basis of proof but on the basis of what they find attractive." Pascal was not talking about people merely believing what they want to believe, as in wish fulfillment. Rather, he was talking about people being swept away by attractive ideas that capture their heart and imagination. Darwinism has played that role for many intellectuals, providing a compelling vision of life and the world.

But visions endure only so long as they can be grounded in reality. The Darwinian vision of life is fast losing touch with reality and specifically with the design that pervades the world at the biochemical level—a world about which Darwin knew nothing. As with all dying paradigms, Darwinism's old guard will not, to paraphrase Dylan Thomas, go gently into that good night. Count on them to rage against the dying light. Notwithstanding, the Darwinian vision is on the way out, to be replaced by a new vision that captures our imagination and at the same time is grounded in reality. Intelligent design is that new vision.

William A. Dembski
Baylor University
Waco, Texas

ACKNOWLEDGMENTS

THIS BOOK OWES MORE TO FOES THAN TO FRIENDS. Just as an oyster gets busy when faced with a challenge, so has it been for me in writing this book. Let me therefore express my gratitude to foes (happily, some of whom are also friends) for devoting an inordinate amount of time, effort and attention to criticizing my work and that of my colleagues in the intelligent design movement. Though sometimes mean-spirited and ill-considered, the criticisms often have been constructive and insightful. Yet invariably I have found them instructive. My hope is that this book, in responding to critics, will likewise prove instructive.

Among the foes, friends and institutions that contributed to this book I want explicitly to thank the following: Dean Anderson; James Barham; Baylor University; Michael Beaty; Michael Behe; David Berlinski; John Bracht; Walter Bradley; J. Budziszewski; Jon Buell; Calvin College; the Center for Theology and the Natural Sciences (CTNS); Bruce Chapman; Robin Collins; Richard Dawkins; Michael Denton; the Discovery Institute's Center for Science and Culture (CSC); Mark Edwards; Wesley Elsberry; Barbara Forrest; the Foundation for Thought and Ethics (FTE); Karl Giberson; Guillermo Gonzalez; Bruce Gordon; Billy Grassie; Paul Gross; Stacy Grote; the International Society for Complexity, Information and Design (ISCID); InterVarsity Press and my editor, Gary Deddo; David Lyle Jeffrey; Phillip Johnson; Steve Jones; Barry Karr; Rob Koons; Gert Korthof; Paul Kurtz; Neil Manson; Nicholas Matzke; Timothy and Lydia McGrew; Angus Menuge; Stephen Meyer; Kenneth Miller; Paul Nelson; Allen Orr; Phylogenists; Massimo Pigliucci; Don Port; Del Ratzsch; Jay Richards; Terry Rickard; Douglas Rudy; Michael Ruse; Andrew Ruys; Donald Schmeltekopf; Thomas Schneider; Eugenie Scott; Michael Shermer; Robert Sloan; Elliott Sober; Micah Sparacio; the Templeton Foundation; Howard Van Till; Richard Wein; Jonathan Wells; John West; John Wilkins; John Wilson; Jonathan Witt and Donald Yerxa. I'm especially grateful to Jonathan Witt for the super job he did editing this manuscript.

Finally, I want to commend my family for always standing by me in my work on intelligent design. Their prayers, encouragement and patience have been an enormous source of strength and comfort to me. Here I want especially to thank my beloved wife, Jana. I also want to thank her parents, John and Dorothy Van Gorp, for their lives of Christian devotion and kindness. I dedicate this book to them.

PART ONE

BASIC DISTINCTIONS

1

INTELLIGENT DESIGN

What is intelligent design?

THINK OF MOUNT RUSHMORE—what about this rock formation convinces us that it was due to a designing intelligence and not merely to wind and erosion? Designed objects like Mount Rushmore exhibit characteristic features or patterns that point us to an intelligence. Such features or patterns are *signs of intelligence.* Proponents of intelligent design, known as design theorists, are not content to regard such signs as mere intuitions. Rather, they insist on studying them formally, rigorously and scientifically.

Intelligent design is the science that studies signs of intelligence. Note that a sign is not the thing signified. Intelligent design does not try to get into the mind of a designer and figure out what a designer is thinking. Its focus is not a designer's mind (the thing signified) but the artifact due to a designer's mind (the sign). What a designer is thinking may be an interesting question, and one may be able to infer something about what a designer is thinking from the designed objects that a designer produces (provided the designer is being honest). But the designer's thought processes lie outside the scope of intelligent design. As a scientific research program, intelligent design investigates the effects of intelligence and not intelligence as such.

What makes intelligent design so controversial is that it purports to find signs of intelligence in biological systems. According to Francisco Ayala, Charles Darwin's greatest achievement was to show how the organized complexity of organisms could be attained without a designing intelligence. Intelligent design therefore directly challenges Darwinism and other naturalistic approaches to the origin and evolution of life. Design has had a turbulent intellectual history. The main difficulty with it in the last two hundred years has been discovering a conceptually powerful formulation of design that will fruitfully advance science. What has kept design outside the scientific mainstream since the rise of Darwinism is that it lacked precise methods for distinguishing intelligently caused objects from unintelligently caused ones.

For design to be a fruitful scientific concept, scientists have to be sure they can reliably determine whether something is designed. For instance, Johannes Kepler thought the craters on the moon were intelligently designed by moon dwellers. We now know that the craters were formed by blind natural processes (like meteor impacts). It's this fear of falsely attributing something to design only to have it overturned later that has prevented design from entering science proper. But design theorists argue that they now have formulated precise methods for discriminating designed from undesigned objects. These methods, they contend, enable them to avoid Kepler's mistake and reliably locate design in biological systems.

As a theory of biological origins and development, intelligent design's central claim is that only intelligent causes adequately explain the complex, information-rich structures of biology and that these causes are empirically detectable. To say intelligent causes are empirically detectable is to say there exist well-defined methods that, based on observable features of the world, can reliably distinguish intelligent causes from undirected natural causes. Many special sciences have already developed such methods for drawing this distinction—notably, forensic science, cryptography, archeology and the search for extraterrestrial intelligence (SETI). Essential to all these methods is the ability to eliminate chance and necessity.

Astronomer Carl Sagan wrote a novel about SETI called *Contact*, which was later made into a movie starring Jodie Foster. The plot and the extraterrestrials, of course, were fictional, but Sagan based the SETI astronomers' methods of design detection squarely on scientific practice. In other words, real-life SETI researchers have never detected designed signals from distant space, but if they encountered such a signal, as the film's astronomers did, they too would infer design. Why did the radio astronomers in *Contact* draw such a design inference from the beeps and pauses they monitored from space? SETI researchers run signals collected from distant space through computers programmed to recognize preset patterns. Signals that do not match any of the patterns pass through the "sieve" and are classified as random.

After years of receiving apparently meaningless "random" signals, the *Contact* researchers discovered a pattern of beats and pauses that corresponds to the sequence of all the prime numbers between 2 and 101. (Prime numbers are divisible only by themselves and by one.) That grabbed their attention, and they immediately detected intelligent design. When a sequence begins with two beats and then a pause, three beats and then a pause, and continues through each prime number all the way to 101 beats, researchers must infer the presence of an extraterrestrial intelligence.

Here's why. Nothing in the laws of physics requires radio signals to take one form or another, so the prime sequence is *contingent* rather than necessary. Also, the prime sequence is a long sequence and therefore *complex*. Note that if the sequence lacked complexity, it could easily have happened by chance. Finally, it was not just complex, but it also exhibited an independently given pattern or *specification*. (It was not just any old sequence of numbers but a mathematically significant one—the prime numbers.)

Intelligence leaves behind a characteristic trademark or signature— what I call *specified complexity*. An event exhibits specified complexity if it is contingent and therefore not necessary; if it is complex and therefore not readily repeatable by chance; and if it is specified in the sense of exhibiting an independently given pattern. Note that a merely improbable event is not sufficient to eliminate chance: flip a coin long enough and you'll witness a highly complex or improbable event. Even so, you'll have no reason not to attribute it to chance.

The important thing about specifications is that they be objectively given and not just imposed on events after the fact. For instance, if an archer fires arrows into a wall and then we paint bull's-eyes around them, we impose a pattern after the fact. On the other hand, if the targets are set up in advance ("specified") and then the archer hits them accurately, we know it was by design.

In determining whether biological organisms exhibit specified complexity, design theorists focus on identifiable systems—such as individual enzymes, metabolic pathways, molecular machines and the like. These systems are specified in virtue of their independent functional requirements, and they exhibit a high degree of complexity. Of course, once an essential part of an organism exhibits specified complexity, then any design attributable to that part carries over to the organism as a whole. One need not demonstrate that every aspect of the organism was designed; in fact, some aspects will be the result of purely natural causes.

The combination of complexity and specification convincingly pointed the radio astronomers in the movie *Contact* to an extraterrestrial intelligence. Specified complexity is the characteristic trademark or signature of intelligence. It is a reliable empirical marker of intelligence in the same way that fingerprints are a reliable empirical marker of a person's presence. Design theorists contend that blind natural causes cannot generate

specified complexity. (See parts two and three. For a full theoretical justification see my 2002 book *No Free Lunch.*)

This isn't to say that naturally occurring systems cannot exhibit specified complexity or that natural processes cannot serve as a conduit for
specified complexity. Naturally occurring systems can exhibit specified
complexity, and nature operating without intelligent direction can take
preexisting specified complexity and shuffle it around. But that is not the
point. The point is whether nature (conceived as a closed system of blind,
unbroken natural causes) can *generate* specified complexity in the sense of
originating it when previously there was none.

Take, for instance, a Dürer woodcut. It arose by mechanically impress-
ing an inked woodblock on paper. The Dürer woodcut exhibits specified
complexity. But the mechanical application of ink to paper via a woodblock does not account for the woodcut's specified complexity. The specified complexity in the woodcut must be referred back to the specified complexity in the woodblock, which in turn must be referred back to the
designing activity of Dürer himself (in this case, deliberately chiseling the
woodblock). Specified complexity's causal chains end not with blind nature but with a designing intelligence.

Biochemist Michael Behe connects specified complexity to biological
design with his concept of *irreducible complexity* (*Darwin's Black Box,* 1996).
Behe defines a system as *irreducibly complex* if it consists of several interrelated parts for which removing even one part completely destroys the system's function. For Behe, irreducible complexity is a sure indicator of design. One irreducibly complex biochemical system that Behe considers is
the bacterial flagellum. The flagellum is an acid-powered rotary motor
with a whiplike tail that spins at twenty thousand revolutions per minute
and whose rotating motion enables a bacterium to navigate through its
watery environment.

Behe shows that the intricate machinery in this molecular motor—including a rotor, a stator, O-rings, bushings and a drive shaft—requires the
coordinated interaction of at least thirty complex proteins and that the absence of any one of these proteins would result in the complete loss of motor function. Behe argues that the Darwinian mechanism faces grave obstacles in trying to account for such irreducibly complex systems. In *No
Free Lunch,* I show how Behe's notion of irreducible complexity constitutes
a special case of specified complexity and that irreducibly complex systems like the bacterial flagellum are therefore designed.

It follows that intelligent design is more than simply the latest in a long line of design arguments. The related concepts of irreducible complexity and specified complexity render intelligent causes empirically detectable and make intelligent design a full-fledged scientific theory, distinguishing it from the design arguments of philosophers and theologians, or what has traditionally been called *natural theology*. According to intelligent design, the world contains events, objects and structures that exhaust the explanatory resources of undirected natural causes and can be adequately explained only by recourse to intelligent causes. Intelligent design demonstrates this rigorously. It thus takes a long-standing philosophical intuition and cashes it out as a scientific research program. This program depends on advances in probability theory, computer science, molecular biology, the philosophy of science and the concept of information—to name but a few. Whether this program can turn design into an effective conceptual tool for investigating and understanding the natural world is for now the big question confronting science.

2

CREATION

How does intelligent design differ from a theological doctrine of creation?

CREATION IS ALWAYS ABOUT THE SOURCE OF BEING OF THE WORLD. Intelligent design is about arrangements of preexisting materials that point to a designing intelligence. Creation and intelligent design are therefore quite different. One can have creation without intelligent design and intelligent design without creation. For instance, one can have a doctrine of creation in which God creates the world in such a way that nothing about the world points to design. Richard Dawkins has a book titled *The Blind Watchmaker: Why the Evidence of Evolution Reveals a Universe Without Design.* Suppose Dawkins is right about the universe revealing no evidence of design. It would not logically follow that it was not created. It is logically possible that God created a world which provides no evidence of his handiwork. On the other hand, it is logically possible that the world is full of signs of intelligence but was not created. This was the ancient Stoic view, in which the world was eternal and uncreated and yet a rational principle pervaded the world and produced marks of intelligence in it.

There's a joke that clarifies the difference between intelligent design and creation. Scientists come to God and claim they can do everything God can do. "Like what?" asks God. "Like creating human beings," say the scientists. "Show me," says God. The scientists say, "Well, we start with some dust and then—" God interrupts, "Wait a second. Get your own dust." Just as a carpenter must take preexisting wood to form a piece of furniture, so these scientists have to take preexisting dust to form a human being. But where did the dust—the raw materials—come from to make a human being? From stars? And where did stars come from? From the big bang? And where did the big bang come from? From a quantum vacuum fluctuation? And where did that quantum vacuum fluctuation come from? At some point such questions must end. Creation asks for an ultimate resting place of explanation: the source of being of the world. Intelligent design, by contrast, inquires not into the ultimate source of matter and energy but into the cause of their present arrangements, particularly those entities, large and small, that exhibit specified complexity.

Although creation and intelligent design are logically separable (you can have one without the other), many who hold to a doctrine of creation also believe that creation exhibits clear marks of intelligence. Biblical texts used to support the connection between creation and intelligent design include Psalm 19:1 ("The heavens declare the glory of God; / the skies proclaim the work of his hands") and Romans 1:20 ("For since the creation of the world God's invisible qualities—his eternal power and divine nature—have been clearly seen, being understood from what has been made"). Thus, many who hold to a doctrine of creation are also proponents of intelligent design. To many theists it seems perfectly reasonable that a creator would create a world in which the creator's intelligence was made manifest. To be sure, the creator could be a master of stealth who obscures his tracks so that they are undetectable. But theists by and large agree that the natural world exhibits God's intelligence, wisdom and purposes.

How the world exhibits design, however, is a matter of dispute. For proponents of intelligent design, design in the world is empirically detectable—we can know it when we see it, and increasingly what enables us to see it is specified complexity. In contrast to taking this scientific approach to design, one can also take a purely theological approach to it. Accordingly, the world exhibits design only against the backdrop of the religious believer's faith experience and theological worldview. On this view, the religious believer sees design in the world only through the eyes of faith. Attributing design to the world thus becomes a theological gloss or overlay, not a generally accessible fact about the world open to believer and nonbeliever alike.

Thus, many theologians resist intelligent design's fundamental claim that the natural world exhibits objectively discernible design. Why is that? For the theist, any designing agent responsible for the world's design would be either God or an intermediary intelligence created by God (e.g., angels, demons or purposeful processes in nature). Such an intermediary would operate either at God's explicit direction or, at least, with God's permission. In any case, God would ultimately be behind all the design in the world. Thus, for instance, any design evident in complex biological systems would have to be ascribed to God. Intelligent design, if it could be developed as a scientific theory applicable to biology, would therefore have immediate theological implications, especially for divine action.

Theologians by and large agree that God acts in the world. But they widely dispute the nature of that activity and whether any aspect of it is open to empirical inquiry. Theology these days is in the grip of several fads, including a preference for divine inscrutability, an overdeveloped need for theodicy and a theology of nature that rules out divine intervention. (We'll return to these in subsequent chapters.) Consequently, theologians and theologically informed scientists often dismiss intelligent design apart from its scientific merits because it clashes with their preconceptions about divine action.

3

SCIENTIFIC CREATIONISM

*Is intelligent design a cleverly disguised
form of scientific creationism?*

INTELLIGENT DESIGN NEEDS TO BE DISTINGUISHED FROM *creation science,* or
scientific creationism. The most obvious difference is that scientific creationism has prior religious commitments whereas intelligent design does not.
Scientific creationism is committed to two religious presuppositions and
interprets the data of science to fit those presuppositions. Intelligent design, by contrast, has no prior religious commitments and interprets the
data of science on generally accepted scientific principles. In particular, intelligent design does not depend on the biblical account of creation. The
two presuppositions of scientific creationism are as follows:

- There exists a supernatural agent who creates and orders the world.

- The biblical account of creation recorded in Genesis is scientifically accurate.

The supernatural agent presupposed by scientific creationism is usually
understood as the transcendent, personal God of the well-known monotheistic religions, specifically Christianity. This God is said to create the
world out of nothing (i.e., without the use of preexisting materials). Moreover, the sequence of events by which this God creates is said to parallel
the biblical record. By contrast, intelligent design nowhere attempts to
identify the intelligent cause responsible for the design in nature, nor does
it prescribe in advance the sequence of events by which this intelligent
cause had to act.

Besides differing in their presuppositions, intelligent design and scientific creationism differ in their propositional content and method of inquiry. Intelligent design begins with data that scientists observe in the laboratory and nature, identifies in them patterns known to signal intelligent
causes and thereby ascertains whether a phenomenon was designed. For

design theorists, the conclusion of design constitutes an inference from data, not a deduction from religious authority. In addition, the propositional content of intelligent design differs significantly from that of scientific creationism. Scientific creationism is committed to the following propositions:

SC1: There was a sudden creation of the universe, energy and life from nothing.

SC2: Mutations and natural selection are insufficient to bring about the development of all living kinds from a single organism.

SC3: Changes of the originally created kinds of plants and animals occur only within fixed limits.

SC4: There is a separate ancestry for humans and apes.

SC5: The earth's geology can be explained via catastrophism, primarily by the occurrence of a worldwide flood.

SC6: The earth and living kinds had a relatively recent inception (on the order of thousands or tens of thousands of years).

Intelligent design, on the other hand, is committed to the following propositions:

ID1: Specified complexity and irreducible complexity are reliable indicators or hallmarks of design.

ID2: Biological systems exhibit specified complexity and employ irreducibly complex subsystems.

ID3: Naturalistic mechanisms or undirected causes do not suffice to explain the origin of specified complexity or irreducible complexity.

ID4: Therefore, intelligent design constitutes the best explanation for the origin of specified complexity and irreducible complexity in biological systems.

A comparison of these two lists shows that intelligent design and scientific creationism differ markedly in content.

Intelligent design is modest in what it attributes to the designing intelligence responsible for the specified complexity in nature. For instance, design theorists recognize that the nature, moral character and purposes of this intelligence lie beyond the competence of science and must be left to religion and philosophy. Intelligent design, as a scientific theory, is distinct from a theological doctrine of creation. Creation presupposes a creator who originates the world and all its materials. Intelligent design only attempts to explain the arrangement of materials within an already given world. Design theorists argue that certain arrangements of matter, especially in biological systems, clearly signal a designing intelligence.

Besides presupposing a supernatural agent, scientific creationism also presupposes the scientific accuracy of the biblical account of creation. Proponents of scientific creationism treat the opening chapters of Genesis as a scientific text and thus argue for a literal six-day creation, the existence of a historical Adam and Eve, a literal Garden of Eden, a catastrophic worldwide flood and so on. Scientific creationism takes the biblical account of creation in Genesis as its starting point and then attempts to match the data of nature to the biblical account.

Intelligent design, by contrast, starts with the data of nature and from there argues that an intelligent cause is responsible for the specified complexity in nature. Moreover, in making such an argument, intelligent design relies not on narrowly held prior assumptions but on reliable methods developed within the scientific community for discriminating designed from undesigned structures. Scientific creationism's reliance on narrowly held prior assumptions undercuts its status as a scientific theory. Intelligent design's reliance on widely accepted scientific principles, on the other hand, ensures its legitimacy as a scientific theory.

These differences between intelligent design and scientific creationism have significant legal implications for advancing intelligent design in the public square. In formulating its position on scientific creationism in *Edwards* v. *Aguillard,* the Supreme Court cited the District Court in *McLean* v. *Arkansas Board of Education.* According to the Supreme Court, scientific creationism is not just similar to the Genesis account of creation but is in fact identical to it and is parallel to no other creation story. Because scientific creationism corresponds point for point with the creation and flood narratives in Genesis, the Supreme Court found scientific creationism to be a religious doctrine and not a scientific theory.

Intelligent design, by contrast, is free from such charges of religious entanglement. Intelligent design is not scientific creationism cloaked in newer and more sophisticated terminology. Intelligent design shares none of scientific creationism's religious commitments. Scientific creationism describes the origin of the universe, its duration, the mechanisms responsible for geological formations, the limits to evolutionary change and the beginnings of humanity, all the while conforming its account of creation to the first chapters of Genesis. In contrast, intelligent design makes no claims about the origin or duration of the universe, is not committed to flood geology, can accommodate any degree of evolutionary change, does not prejudge how human beings arose and does not specify in advance how a designing intelligence brought the first organisms into being.

Consequently, it is mistaken and unfair to confuse intelligent design with scientific creationism. Intelligent design is a strictly scientific theory devoid of religious commitments. Whereas the creator underlying scientific creationism conforms to a strict, literalist interpretation of the Bible, the designer underlying intelligent design need not even be a deity. To be sure, the designer is compatible with the creator-God of the world's major monotheistic religions, such as Judaism, Christianity and Islam. But the designer is also compatible with the watchmaker-God of the deists, the Demiurge of Plato's *Timaeus* and the divine reason (i.e., *logos spermatikos*) of the ancient Stoics. One can even take an agnostic view about the designer, treating specified complexity as a brute fact inherently unexplainable in terms of chance and necessity. Unlike scientific creationism, intelligent design does not prejudge such questions as Who is the designer? or How does the designer go about designing and building things?

4

DISGUISED THEOLOGY

Even though intelligent design purports to be a scientific research program, isn't it really a theological enterprise?

THE BIG BANG HAS THEOLOGICAL IMPLICATIONS, but that does not make it a theological enterprise. Likewise, intelligent design has theological implications, but that does not make it a theological enterprise. Intelligent design is an emerging scientific research program. Design theorists attempt to demonstrate its merits fair and square in the scientific world—without appealing to religious authority. The fundamental claim of intelligent design is straightforward and easily intelligible: namely, *there exist natural systems that cannot be adequately explained in terms of undirected natural causes and that exhibit features which in any other circumstance we would attribute to intelligence.* That claim can be considered on its own merits. Let's go to nature, identify some natural systems, analyze them and see whether the analysis leads us to design.

Do certain types of natural systems exhibit clear hallmarks of intelligence? This is a perfectly legitimate scientific question. Moreover, the answer to this question cannot be decided on philosophical, theological or ideological grounds but must be decided through careful scientific investigation. To be sure, intelligent design has its work cut out for it, and the required analysis to answer this question is only now beginning. But instead of encouraging a fair scientific assessment of it, critics of intelligent design often do everything in their power to delegitimize this question so that it cannot receive a fair hearing within the scientific community. Rather than help assess the merit of intelligent design as a scientific project, they relegate it to the "safe" realms of religion and theology, where it can't cause any trouble (which in itself is an indictment of how far theology has been downgraded in Western culture).

Why are critics of intelligent design so quick to conflate it with theology—and a disreputable form of theology at that? Darwinists like Kenneth Miller and Robert Pennock, who write full-length books on intelligent design, lament that it is theology masquerading as science. (See Miller's *Finding Darwin's God* and Pennock's *Tower of Babel*, both of which were published in 1999.) To this theologians like John Haught and Ian Barbour add that intelligent design doesn't even succeed as theology. Why is that? Indeed, why does Miller write a book titled *Finding Darwin's God* and why does Haught write a book titled *God After Darwin*? The juxtaposition here of God and Darwin is not coincidental.

I submit that the preoccupation by critics of intelligent design with theology results not from intelligent design being inherently theological. Instead, it results from critics having built their own theology (or anti-theology, as the case may be) on a foundation of Darwinism. Intelligent design challenges that foundation, so critics reflexively assume that intelligent design must be inherently theological and have a theological agenda. Freud, if it were not for his own virulent Darwinism, would have instantly seen this as a projection. Critics of intelligent design resort to a classic defense mechanism: they project onto intelligent design the very thing that intelligent design unmasks in their own views, namely, that Darwinism, especially as it has been taken up by today's intellectual elite, has itself become a project in theology.

Consider Barbour's comments at a meeting of the American Academy of Religion (Nashville, November 19, 2000). At that meeting Barbour claimed that intelligent design is a form of natural theology, a designation that in today's science-religion dialogue guarantees it second-class status. But what is Barbour's alternative to natural theology? He writes, "My own approach is not natural theology but a theology of nature in which one asks how nature as understood by science is related to the divine as understood from the religious experience of a historical community." In offering a theology of nature rather than a natural theology, Barbour purports to capture the intellectual high ground.

But why does Barbour think that? Indeed, why in an address to the American Academy of Religion does Barbour need to stress that design advocates like Huston Smith "underestimate the weight of evidence favoring neo-Darwinian theory"? Why in that same talk does he emphasize that "the scientific account is complete on its own level" and that "scientists have to assume methodological naturalism, that is, they seek explanations in terms of natural causes"? Why does Barbour perpetuate the myth that "the God of the gaps has steadily retreated in the history of modern science" when the history of science is filled with cases where scientists thought they had resolved a problem only to discover they hadn't? In short, why as a theologian is Barbour so concerned about preserving Darwinism and the naturalistic conception of science that undergirds it?

The answer, clearly, is that Barbour has built his "theology of nature," as he calls it, on such a naturalistic foundation. Specifically, Barbour presupposes that nature is a complete system of natural causes and that the Darwinian mechanism is the means by which biological complexity has emerged within nature. As a consequence, intelligent design cannot appear to him as anything but a thoroughly theological enterprise. Yet intelligent design is not a theological enterprise. It only seems like a theological enterprise because, as a scientific theory that challenges Darwinism, intelligent design challenges the theological edifice that Barbour himself has built on Darwinism.

To challenge a foundation is to challenge any edifice built on that foundation. That theological edifice, which Barbour refers to as a theology of nature, is rightly understood as a natural theology. To be sure, it is not a natural theology of the classic "isn't it amazing how your legs are just long enough to reach the ground" variety that the British natural theologians are widely caricatured as having exemplified. (In fact, some of the British natural theologians, like Robert Boyle, were far more subtle than we ordinarily give them credit.) But the basic impulse behind natural theology—to take the science of the day, baptize it and use it for theological mileage—is certainly there in Barbour's work.

A lot of theology and anti-theology has been built on Darwinism. (Intelligent design theorist Cornelius Hunter details just how much so in his book *Darwin's God.*) The anti-theology of Richard Dawkins, Daniel Dennett and William Provine is well known. But the positive theology that gets built on Darwinism is worth exploring because its connection to more traditional theologies is not always clear. In describing his theology of nature, for instance, Barbour characterizes the theologian's task as inquiring "how nature as understood by science is related to the divine as understood from the religious experience of a historical community." Given Barbour's description of his theology of nature, we may ask, what exactly is "the divine as understood from the religious experience of a historical community"? Traditional theologies—whether Jewish, Christian or Muslim—take as their basic datum divine revelation (e.g., God speaking to Moses on Mount Sinai) and view that revelation as encapsulated in inspired and authoritative texts that have an objective sense and that are binding on believers.

To illustrate the difference, consider the analogy of an oil painting. An oil painting is ordinarily painted on a canvas. One can therefore ask whether the canvas is designed. Alternatively, one can ask whether some configuration of paint on the canvas is designed. The design of the canvas corresponds to the design of the universe as a whole. The design of some configuration of paint corresponds to an instance of design within the universe.

In this analogy, the universe is a canvas on which is depicted natural history. One can ask whether that canvas is itself designed. On the other hand, one can ask whether features of natural history depicted on that canvas are designed. In biology, for instance, one can ask whether an irreducibly complex biochemical machine like the bacterial flagellum is designed. Although design remains a much discussed topic in cosmology (is the universe as a whole designed?), with intelligent design's focus on biology, most of the discussion and controversy now centers on biology (is there design in the universe and, specifically, in biology?).

But divine revelation is not the decisive factor for Barbour and others who build their theology on the deliverances of science. Instead, the decisive factor is how the divine is "understood from the religious experience of a historical community." All the emphasis here is on the understanding of the religious community and not on the divine self-revelation that within traditional theologies is the reason for those communities arising in the first place. Ultimately what's decisive for Barbour is how the community as it has come down to the present day understands its religious experience.

Now I don't mean to suggest that this source of theological reflection is irrelevant in what I'm calling traditional theology. But in emphasizing our current understanding of religious experience as opposed to our obligation to align ourselves with an objective revelation, Barbour opens the door to radical re-understandings of the divine as the religious experience of the community of faith evolves. And evolve it has, especially in the light of Darwinism. Once Darwinism conditions religious experience, theology experiences an irresistible urge to universalize evolution as a principle that applies even to the divine. Thus the unchanging God of traditional theologies gives way to the evolving God of process theologies. Thus traditional theism with its strong transcendence gives way to panentheism, with its modified transcendence wherein God is inseparable from and dependent on the world.

Let me stress that I'm not arguing here for the superiority of one approach to theology over another (though I certainly have my own preferences). My point is simply that Darwinism has radical implications for theology and that in challenging Darwinism, intelligent design likewise has radical implications for theology. This is not to say that intelligent design is a theological enterprise any more than Darwinism is a theological enterprise. Darwinism, conceived as a theory about how biological complexity has emerged in the history of life, is a scientific theory. Intelligent design, conceived as a theory about the inherent limitations of undirected natural causes to generate biological complexity and the need for intelligence to overcome those limitations, is likewise a scientific theory.

It should be no surprise that intelligent design is as controversial as it is. Intelligent design doesn't merely challenge the high priests of Darwinism. It also highlights the breach between popular culture, which is largely committed to intelligent design, and high culture, which largely rejects it in favor of Darwinian naturalism. Our intuitions invariably begin with design. Only by being suitably educated (indoctrinated) are we educated out of those intuitions. Even the arch-skeptic Michael Shermer admits as much in his book *How We Believe.* People in North America overwhelmingly believe in God. According to a poll of ten thousand people that Shermer commissioned, the top reason people believe in God is the order and complexity they observe in the natural world and the evidence these are supposed to provide for design.

The problem to date is that our common intuitions about design have been inchoate, pretheoretical and theological. On the other hand, our reasons for rejecting design as a result of Darwinism have been extensively developed, thoroughly advertised and without apparent theological precommitments. Intelligent design is turning the tables on this disparity by placing those inchoate and pretheoretical intuitions of design on a firm rational foundation and by carefully distinguishing design from theology (especially from natural theology).

Darwinists, who have grown accustomed to holding the intellectual high ground, are understandably reluctant to relinquish their monopoly over high culture. The question is whether they will continue to misrepresent intelligent design as a theological enterprise to artificially insulate their theory from competition, or whether they will step onto the moral high ground by opening scientific discussions to the questions intelligent design raises. Not having a particularly optimistic view of human nature, I expect Darwinists will continue business as usual, misrepresenting intelligent design as long as they can get away with it and clinging to their monopoly over biological education as long as a cowed public will permit them. My hope for the success of intelligent design, therefore, resides not with Darwinists but with a younger generation of scholars who can dispassionately consider the competing claims of Darwinism and intelligent design.

5

RELIGIOUS MOTIVATION

*Isn't the real driving force behind
intelligent design a fear that evolutionary
theories, and Darwinism in particular,
will one day permanently displace any need
for God?*

ACCORDING TO SOME CRITICS OF INTELLIGENT DESIGN, design theorists oppose Darwinism not because of a concern for truth but because of a deep-seated fear that Darwinism destroys traditional morality and religious belief. Such critics find it inconceivable that someone, once properly exposed to Darwin's theory, could fail to embrace it wholeheartedly, much less doubt it. It is as though Darwin's theory were one of René Descartes's clear and distinct ideas that immediately impel assent. Thus for design theorists to oppose Darwin's theory requires some hidden motivation, like wanting to shore up traditional morality or being a closet fundamentalist.

For the record, therefore, let's be clear that design theorists oppose Darwinian theory on strictly scientific grounds. Yes, we are interested in and write about the theological and cultural implications of Darwinism's imminent demise and replacement by intelligent design. But the reason design theorists take seriously such implications is that we are convinced that Darwinism is, on its own terms, an oversold and overreaching scientific theory.

Darwinism has achieved the status of inviolable science. As a consequence, design theorists encounter a ruthless dogmatism when challenging Darwin's theory. The problem isn't that Darwinists don't hold their theory tentatively. No scientist with a career invested in a scientific theory is going to relinquish it easily. By itself a scientist's lack of tentativeness poses no danger to science. It only becomes a danger when it turns to dogmatism. Typically, a scientist's lack of tentativeness toward a scientific theory simply means that the scientist is convinced the theory is substantially correct. Scientists are fully entitled to such convictions. On the other hand, scientists who hold their theories dogmatically go on to assert that their theories *cannot* be incorrect. Moreover, scientists who are ruthless in their dogmatism regard their theories as inviolable and portray critics as morally and intellectually deficient.

How can a scientist keep from descending into dogmatism? The only way I know is to look oneself squarely in the mirror and continually affirm, *I am a fallible human being. I may be wrong. I may be massively wrong. I may be hopelessly and irretrievably wrong*—and mean it! It's not enough just to mouth these words. We need to take them seriously and admit that they can apply even to our most cherished scientific beliefs. (This injunction holds true as much for design theorists as for Darwinists.) Human fallibility is real and can catch us in the most unexpected places.

A simple induction from past scientific failures should be enough to convince us that the only thing about which we cannot be wrong is the possibility that we might be wrong. Such a self-deflating skepticism cuts deeper than any Cartesian skepticism, which always admitted some privileged domains of knowledge that were immune to doubt. (For Descartes, mathematics and theology constituted such domains.) It also cuts deeper than the selective skepticism of today's professional skeptics, for whom Darwinism and naturalism are nonnegotiables (see chapter twenty-eight). At the same time, this self-deflating skepticism is consonant with an abiding faith in human inquiry and its ability to render the world intelligible. Indeed, the conviction with which scientists hold their scientific theories, so long as it is free of dogmatism, is just another word for faith. This faith sees the scientific enterprise as fundamentally worthwhile even if any of its particular claims and theories might be overturned. This is not an arbitrary but a circumspect faith. It takes the possibility of error not as an obstacle to knowledge but as a basis for humility, a humility that is always willing to question and test to make sure we are not deceiving ourselves. It is a faith that leaves room for doubt.

In place of such a faith, dogmatism substitutes unreasoning certainty in particular claims and theories of science. The problem with dogmatism is that it is always a form of self-deception. If Socrates taught us anything, it's that we always know a lot less than we think we know. Dogmatism deceives us into thinking we have attained ultimate mastery and that divergence of opinion is futile. Self-deception is the original sin because it deceives us into believing that self-deception is impossible. Richard Feynman put it this way: "The first principle is that you must not fool yourself, and you are the easiest person to fool." Feynman was particularly concerned about applying this principle to the public understanding of science: "You should not fool the laymen when you're talking as a scientist. . . . I'm talking about a specific, extra type of integrity that is [more than] not lying, but bending over backwards to show how you're maybe wrong." (See Feynman's autobiography, *"Surely You're Joking, Mr. Feynman!"*)

Sadly, Feynman's sound advice almost invariably gets lost when Darwin's theory is challenged. It hardly makes for a free and open exchange of ideas when biologist Richard Dawkins asserts, "It is absolutely safe to say that if you meet somebody who claims not to believe in evolution, that person is ignorant, stupid or insane (or wicked, but I'd rather not consider that)." (Dawkins made this remark in his 1989 *New York Times* review of Donald Johanson and Maitland Edey's book *Blueprints*.) Nor does philosopher Daniel Dennett help matters when, in *Darwin's Dangerous Idea*, he recommends quarantining religious parents who object to their children being taught evolutionary theory. Nor for that matter does skeptic Michael Shermer promote insight into the Darwinian mechanism of natural selection when, in *Why People Believe Weird Things*, he announces, "No one, and I mean *no one*, working in the field is debating whether natural selection is the driving force behind evolution, much less whether evolution happened or not."

Such remarks, and the arrogance they betray, do nothing to ameliorate the ongoing controversy over Darwinian evolution. Gallup polls consistently indicate that only about 10 percent of the U.S. population accepts the sort of evolution advocated by Dawkins, Dennett and Shermer—that is, evolution in which the driving force is the Darwinian selection mechanism or some other purely natural mechanism. The rest of the population is committed to some form of intelligent design. Calling the majority of Americans names or wishing to quarantine those who are not sufficiently sympathetic to Darwinism isn't how science moves forward. It's not how a scientific theory wins adherents on the merits of its evidence.

Now it goes without saying that science is not decided in an opinion poll. Nevertheless, the overwhelming rejection of Darwinian evolution in the population at large is worth pondering. Although Shermer exaggerates when he claims that no research biologist doubts the power of natural selection, he would certainly be right to claim that this is the majority position among biologists. Why, then, has the biological community failed to convince the public that natural selection is the driving force behind evolution and that evolution so conceived (i.e., Darwinian evolution) can successfully account for the full diversity of life? This question is worth pondering because in most other areas of science, the public prefers to sign off on the considered judgments of the scientific community. (Science, after all, holds considerable prestige in our culture.) Why is this not the case here? Steeped as our culture is in the fundamentalist-modernist controversy, the usual answer is that religious fundamentalists, blinded by their dogmatic prejudices, willfully refuse to acknowledge the overwhelming case for Darwinian evolution.

The problem with this explanation is that fundamentalism, in the sense of strict biblical literalism, is a minority position among religious believers. Most religious traditions do not make a virtue out of alienating the culture. Despite postmodernity's inroads, science retains tremendous cultural prestige. The religious world by and large would rather live in harmony with the scientific world. Most religious believers accept that species have undergone significant changes over the course of natural history and, therefore, that evolution has in some sense occurred. (Consider, for instance, Pope John Paul II's 1996 endorsement of evolution in his address to the Pontifical Academy of Sciences titled "Truth Cannot Contradict Truth.") The question for religious believers and the public more generally is the extent of evolutionary change and the mechanism underlying evolutionary change—in particular, whether material mechanisms alone are sufficient to explain all of life. In short, the real reason the public continues to resist Darwinian evolution is that the Darwinian mechanism of chance variation and natural selection seems inadequate to account for the full diversity of life.

One frequently gets the sense from reading publications by the National Academy of Sciences, the National Center for Science Education and the National Association of Biology Teachers that the public's failure to accept Darwinian evolution is a failure in education. That is, if only people could be made to understand Darwin's theory properly, so we are told, they would readily sign off on it. What an odd assumption, given that Darwinists hold a virtual monopoly on biology education in America. Accordingly, a mindless fundamentalism must reign over the minds of a vast majority of Americans. For what else could prevent Darwinism's immediate and cheerful acceptance except religious prejudice?

Thus, what many Darwinists yearn for is not just more talented communicators to promote Darwinism in America's biology classrooms but an enforced educational and cultural policy for total worldview reprogramming that is sufficiently aggressive to capture and convert to Darwinism even the most recalcitrant among "religiously programmed" youth. That's why Darwinists like Dennett—by all appearances a functioning member and advocate of democracy—fantasizes about quarantining religious parents. It seems ridiculous to convinced Darwinists like him that the fault might lie with their theory and that the public might be picking up on faults inherent in their theory. And yet that is exactly what is happening.

The public need feel no shame at disbelieving and openly criticizing Darwinism. Most scientific theories these days are initially published in specialized journals or monographs and are directed toward experts who, it is assumed, possess considerable technical background and competence (see chapter forty-one). Not so Darwin's theory. The locus classicus for Darwin's theory remains his *Origin of Species*. In it Darwin took his case to the public. Contemporary Darwinists likewise continue to take their case to the public. Books by Richard Dawkins, Daniel Dennett, E. O. Wilson, the late Stephen Jay Gould and a host of other biologists and philosophers aim to convince a skeptical public of the merits of Darwin's theory. These same authors commend the public when it finds their arguments convincing. But when the public remains unconvinced, commendation turns to condemnation and even vilification. The mark of dogmatism is to reward conformity and punish dissent. If contemporary science does indeed belong to the culture of rational discourse, then it must repudiate dogmatism and authoritarianism in all guises.

Why does the public find the case for Darwinism unconvincing? Fundamentalism aside, the claim that the Darwinian mechanism of chance variation and natural selection can generate the full range of biological diversity strikes people as an unwarranted extrapolation from the limited changes that mechanism is known to effect in practice. The hard empirical evidence for the power of the Darwinian mechanism is in fact quite limited (e.g., finch beak variation, changes in flower coloration, bacteria developing antibiotic resistance). For instance, finch beak size does vary according to environmental pressure. The Darwinian mechanism does operate here and accounts for the changes we observe. But that same Darwinian mechanism is also supposed to account for how finches arose in the first place. This is an extrapolation. Strict Darwinists see it as perfectly plausible. The public remains unconvinced.

But shouldn't the public simply defer to the scientists? After all, they are the experts. But to which scientists should they defer? It's certainly the case that the majority of the scientific community accepts Darwinism. But science is not decided at the ballot box, and Darwinism's acceptance among scientists is hardly universal. The theory of intelligent design is rapidly gaining advocates at the highest level of the academy, both in the humanities and in the sciences. (To see this, refer to <www.iscid.org>, the website of the International Society for Complexity, Information and Design.) Whether intelligent design ultimately overturns Darwinism, however, is not the issue here. The issue is whether the scientific community is willing to eschew dogmatism and admit as a live possibility that even its most cherished views might be wrong.

Darwinists will have none of this. Instead, they resort to the circular logic of defining true scientists as only those who accept Darwinism. Having defined their scientific opposition out of existence, they then make the historically dubious claim that when a scientific community universally supports a position, it must be correct. Such machinations are unworthy of science. Scientists have been wrong in the past and will continue to be wrong, both about niggling details and about broad conceptual matters. Darwinism is one scientific theory that attempts to account for the history of life; but it is not the only scientific theory that could possibly account for it. It is a widely disputed theory, one that is facing ever more trenchant criticisms and, like any other scientific theory, needs periodic reality checks.

One reality check is to determine whether intelligent design may legitimately be taught in the public school science curriculum. Opponents of intelligent design try to argue that because many of its proponents are religious believers who want to see intelligent design prosper as a way to cultural and scientific renewal, intelligent design is therefore religiously motivated and may not legitimately be taught in public schools. According to the Lemon Test, which was first enunciated by the Supreme Court in *Lemon* v. *Kurtzman* and which specifies whether something is religious for purposes of the Establishment Clause, the curriculum of public schools must have a secular purpose. Intelligent design, by being religiously motivated, is said to have a religious purpose and must therefore be kept outside the public school science curriculum.

The problem with this argument is that it conflates motivation and purpose. The distinction between motivation and purpose is well understood in the criminal law context but typically gets lost in discussions about intelligent design. If you enter your mother's nursing home and smother her with a pillow, the law is not interested in your *motive*. Was it to gain your inheritance more quickly, to settle a long-standing grudge or to comply compassionately and tearfully with her request to put her out of her misery? The law doesn't care about such motives (at least not principally).

Whether you are convicted of murder depends on whether you had the *purpose* to end her life.

In this distinction, motivation refers to what moves us to act whereas purpose refers to how we channel or direct our actions in response to our motives. Purpose gives expression to motivation. It follows that there need be no correlation between the validity of motives and the validity of purposes. One might have good motives but be wicked at implementing them and thus have evil purposes. For instance, motivated by the desire to stop urban violence, someone might become a vigilante. On the other hand, one can have evil motives but attempt to realize them through purposes that (happily) produce more good than harm. For instance, motivated by hatred and fear of an ethnic minority coworker, one might arrange to have another firm hire the employee for a better paying and better suited job.

Thus, whenever the National Center for Science Education, the American Civil Liberties Union and other such organizations assert that intelligent design is religiously motivated and therefore doesn't deserve the same respect as other ideas that may legitimately be discussed in the academy, we need to distinguish clearly between motivation and purpose. So long as intelligent design has a demonstrable secular purpose—advancing science, enriching the science curriculum, preventing viewpoint discrimination, promoting academic freedom—its motivation, even if religious, is legally irrelevant.

6

OPTIMAL DESIGN

Why place the word intelligent *in front of* design? *It seems that much of the design in nature is anything but intelligent.*

THE WORD *INTELLIGENT* HAS TWO MEANINGS. It can simply refer to the activity of an intelligent agent, even one that acts stupidly. On the other hand, it can mean that an intelligent agent acted with skill and mastery. Failure to draw this distinction results in confusion about intelligent design. This was brought home to me in a radio interview. Skeptic Michael Shermer and paleontologist Donald Prothero were interviewing me on National Public Radio. As the discussion unfolded, I was surprised to find that how they used the phrase "intelligent design" differed significantly from how the intelligent design community uses it.

Shermer and Prothero understood the word *intelligent* in "intelligent design" in the sense of clever or masterful design. They therefore presumed that intelligent design must entail optimal design. The intelligent design community, on the other hand, understands the *intelligent* in "intelligent design" simply as referring to intelligent agency (irrespective of skill or mastery) and thus separates intelligent design from optimality of design.

But why then place the adjective *intelligent* in front of the noun *design*? Doesn't the concept of design already include the idea of intelligent agency, so that juxtaposing the two becomes redundant? Redundancy is avoided because intelligent design needs to be distinguished from apparent design on the one hand and optimal design on the other. Intelligent design stresses that the design is due to an actual intelligence, but it leaves entirely open the attributes or qualities of that intelligence.

Apparent design, by contrast, asserts that the design is not actual. For instance, Richard Dawkins begins his book *The Blind Watchmaker* with the quotation, "Biology is the study of complicated things that give the appearance of having been designed for a purpose." Dawkins then requires an additional three hundred pages to argue that this design is only an appearance and is not actual. Apparent design therefore constitutes a negation of intelligent design.

Many biologists sidestep intelligent design and the evidence for it by shuttling between apparent design and optimal design. To argue for apparent design, they simply lay out the case for pure, unaided Darwinism. To argue against intelligent design, they substitute a handy strawman, identifying intelligent design with optimal design. To render intelligent design as implausible as possible, they then define optimal design as perfect design that is best with respect to every possible criterion of optimization. (Anything less, presumably, would not be worthy of an intelligent designer.) Since actual designs always involve tradeoffs and compromise, such globally-optimal-in-every-respect designs cannot exist except in an idealized realm (sometimes called a "Platonic heaven") far removed from the actual designs of this world. Unlike intelligent design, apparent design and optimal design empty design of practical significance.

Assimilating all biological design to either apparent or optimal design avoids the central question that needs to be answered, namely, whether there actually is design in biological systems regardless of what additional attributes they possess (like optimality). The automobiles that roll off the assembly plants in Detroit are intelligently designed in the sense that actual human intelligences are responsible for them. Nevertheless, even if we think Detroit manufactures the best cars in the world, it would still be wrong to say that they are optimally designed. Nor would it be correct to say that they are only apparently designed (and certainly not for the reason that they fail to be optimally designed). Is there an even minimally sensible reason for insisting that design theorists must demonstrate optimal design in nature? Critics of intelligent design (e.g., the late Stephen Jay Gould) often suggest that any purported cosmic designer would only design optimally. But that is a theological rather than a scientific claim.

Although attributing intelligent design to human artifacts is uncontroversial, eyebrows are quickly raised when intelligent design is attributed to biological systems. Applied to biology, intelligent design maintains that a designing intelligence is required to account for the complex, information-rich structures in living systems. At the same time, it refuses to speculate about the nature of that designing intelligence. Whereas optimal design demands a perfectionistic designer who has to get everything just right, intelligent design fits our ordinary experience of design, which is conditioned by the needs of a situation, requires negotiation and tradeoffs, and therefore always falls short of some idealized global optimum.

No real designer attempts optimality in the sense of attaining perfect design. Indeed, there is no such thing as perfect design. Real designers strive for *constrained optimization*, which is something altogether different. As Henry Petroski, an engineer and historian at Duke University, aptly remarks in *Invention by Design*, "All design involves conflicting objectives and hence compromise, and the best designs will always be those that come up with the best compromise." Constrained optimization is the art of compromise among conflicting objectives. This is what design is all about. To find fault with biological design because it misses some idealized optimum, as Gould regularly used to do, is simply gratuitous. Not knowing the objectives of the designer, Gould was in no position to say whether the designer proposed a faulty compromise among those objectives.

Nonetheless, the claim that biological design is suboptimal has been tremendously successful at shutting down discussion about design. Interestingly, that success comes not from analyzing a given biological structure and showing how a constrained optimization for constructing that structure might have been improved. That would constitute a legitimate scientific inquiry, so long as the proposed improvements could be concretely implemented and did not degenerate into wish-fulfillment, where one imagines some improvement but has no idea how it can be effected or whether it might lead to deficits elsewhere. Just because we can always imagine some improvement in design doesn't mean that the structure in question wasn't designed, or that the improvement can be effected, or that the improvement, even if it could be effected, would not entail deficits elsewhere.

In my public lectures, I'm frequently asked about the alleged suboptimal design of the human organism. Among the things a putative designer of the human organism is charged with botching are the convergence of larynx and esophagus at the pharynx and the resulting susceptibility to choking; poor back construction and the resulting back pain and loss of mobility; faulty pelvis construction in females and the resulting difficulty giving birth; and smallness of the human jaw and its resultant inability to accommodate the full set of "primate teeth" (the most notable consequence being difficulties with wisdom teeth). At the top of the list, however, is the topsy-turvy arrangement of the human eye. The problem with the human eye, evolutionary biologists endlessly tell us, is that it has an inverted retina. Accordingly, the photoreceptors in the eye are oriented away from incoming light and situated behind nerves and blood vessels, which are said to obstruct the incoming light.

In fact, there appear to be good functional reasons for this construction. A visual system needs three things: speed, sensitivity and resolution. Speed is unaffected by the inverse wiring. Resolution seems unaffected as well (save for a tiny blind spot, which the brain seems to work around without difficulty). Indeed, there is no evidence that the cephalopod retina of squids and octopuses, which is "correctly wired" by having receptors facing forward and nerves tucked behind, is any better at resolving objects in its visual field. As for sensitivity, however, it seems that there are good functional reasons for an inverted retina. Retinal cells require the most oxygen of any cells in the human body. But when do they require the most oxygen? Their oxygen requirement is maximal when incident light is minimal. Having a blood supply in front of the photoreceptors guarantees that the retinal cells will have the oxygen they need to be as sensitive as possible when incident light is minimal. (Some vertebrate eyes with inverted retinas are so sensitive that they can respond to single photons.)

Now my point here is not that the human eye can't be improved or is in some ultimate sense optimal. My point, rather, is that simply drawing attention to the inverted retina is not a reason to think that eyes with that structure are suboptimal. Indeed, there are no concrete proposals on the table for how the human eye might be improved that can also guarantee no loss in speed, sensitivity and resolution. There's also an irony here worth noting: the very visual system that is supposed to be so poorly designed and that no self-respecting designer would have constructed is nonetheless good enough to tell us that the eye is inferior. We study the eye by means of the eye. And yet the information that the eye gives us is supposed to show that the eye is inferior. This is one of many cases in evolutionary biology in which scientists bite the hand that feeds them.

Design is a matter of tradeoffs. There's no question that we would like to add to or improve existing designs by conferring additional functionalities. It would be nice to have all the functionality of the human eye without a blind spot. It would be nice to have all the functionality of the respiratory and food-intake system as well as a reduced incidence of choking. It would be nice to have all the functionality of our backs and a decreased incidence of back pain. It would be nice to have all the functionality of the female pelvis along with easier delivery of children. It would be nice to have all the functionality of our jaws without wisdom teeth. But when the suboptimality objection is raised, invariably one finds only additional functionalities mentioned but no details about how they might be implemented. And with design, the devil is in the details.

Yet even if such details were forthcoming, they would undercut not design as such but only its quality (i.e., its degree of excellence). And even here we have to be careful. Just because a design could be improved in the sense of increasing the functionality of some aspect of an organism, this does not mean that such an improvement would be beneficial within the wider ecosystem within which the organism finds itself. A functionality belonging to a predator might be vastly improvable, but it also might render the predator that much more dangerous to its prey and thereby drastically alter the balance of the ecosystem, conceivably to the detriment of the entire ecosystem. In criticizing design, biologists tend to place a premium on functionalities of individual organisms and see design as optimal to the degree that those individual functionalities are maximized. But higher-order designs of entire ecosystems might require lower-order designs of individual organisms to fall short of maximal function.

Our view of design is shaped too much by sports competitions. We always want to go faster, higher, longer and stronger. But do we really want to go faster, higher, longer and stronger without limit? Of course not. It is precisely the limits on functionalities that make the game of life interesting. (That's why many games employ handicaps.) A five-hundred-pound, seven-foot-six football player with the strength of a gorilla and the speed of a cheetah would instantly be banned from the sport, because just by playing the game to the best of one's ability, such a player would maim or kill all normal players who got in the way.

Fans might show up to such a game for the novelty of it or out of bloodlust, but a player like this would destroy the competitive drama of the game. Indeed, before long this super-player would destroy or run off anyone willing to play the game, and there would cease to be a game. Likewise, such a predator in an ecosystem would wipe out all the prey, after which it would go extinct. Or if the super-creature were omnivorous, it would reproduce optimally (like rabbits? like bacteria?) until it wiped out all life, after which it would again go extinct (unless it became an autotroph and could manufacture its food from scratch as do some single-celled organisms).

Biology is, among other things, a drama. Interesting dramas require characters who are less than optimal in some respects. In fact, authors of human dramas often consciously design their characters with flaws and weaknesses. Would *Hamlet* be nearly as interesting if Shakespeare had not designed the play's lead character to exhibit certain flaws and weaknesses, notably indecisiveness?

I'm not saying that weaknesses or flaws in the design characteristics of organisms or ecosystems can be the basis for a design inference. Design inferences are drawn by identifying features of systems that are uniquely diagnostic of intelligence. At the same time, weaknesses or flaws in the design characteristics of organisms or ecosystems could be compatible with evolutionary changes guided by an intelligence. Such an evolutionary scenario—in which not every aspect of organisms taken in isolation is optimal—would not entail that any intelligence guiding evolutionary change is necessarily flawed.

Critics of intelligent design repeatedly claim that no expert designer would have created all the evolutionary dead-ends we see in the fossil record. One of my critics asks, "What might be the intelligent purpose for creating species doomed for extinction? Or why would an intelligent designer create humans with spines poorly adapted for bipedal locomotion?" If we think of evolution as progressive in the sense that the capacities of organisms get honed and false starts get weeded out by natural selection over time, then it seems implausible that a wise and benevolent designer might want to guide such a process. But if we think of evolution as regressive, as reflecting a distorted moral structure that takes human rebellion against the designer as a starting point, then it's possible a flawless designer might use a very imperfect evolutionary process as a means of bringing a prodigal universe back to its senses. But this is an idea to be explored in another book.

We've veered a long way from science, and for good reason. In arguing that nature couldn't be designed because various biological systems are suboptimal, opponents of intelligent design have shifted the terms of the discussion from science to theology. In place of, How specifically can an existing structure be improved? the question instead becomes, Would any self-respecting deity really create a structure like that? Gould was a master of this bait-and-switch. For instance, in *The Panda's Thumb* he wrote:

> If God had designed a beautiful machine to reflect his wisdom and power, surely he would not have used a collection of parts generally fashioned for other purposes. . . . Odd arrangements and funny solutions are the proof of evolution—paths that a sensible God would never tread but that a natural process, constrained by history, follows perforce.

Gould was criticizing here what's called the panda's "thumb," a bony extrusion that helps the panda strip bamboo of its hard exterior and thus render the bamboo edible to the panda. (The panda's thumb, which is an enlarged radial sesamoid, in fact serves the panda extremely well in rendering bamboo edible.)

The first question that needs to be answered about the panda's thumb, and indeed about any biological structure, is whether it displays clear marks of intelligence. The design theorist is not committed to every biological structure being designed. Naturalistic mechanisms like mutation and selection do operate in natural history to adapt organisms to their environments. Perhaps the panda's thumb is such an adaptation. Nonetheless, naturalistic mechanisms are incapable of generating the highly specific, information-rich structures that pervade biology. Organisms display the hallmarks of intelligently engineered high-tech systems—information storage and transfer, functioning codes, sorting and delivery systems, self-regulation and feedback loops, signal-transduction circuitry—and everywhere, complex arrangements of mutually interdependent and well-fitted parts that work in concert to perform a function. Opponents of intelligent design are fond of equivocating, staging ad hominem attacks, slaying strawmen, making simplistic theological claims in the guise of science or simply stonewalling. What they are not fond of is squarely facing the astonishing evidence for intelligent design and seeking to refute it point by logical point.

7

THE DESIGN ARGUMENT

How does intelligent design differ from the design argument?

THE DESIGN ARGUMENT BEGINS WITH features of the physical world that exhibit evidence of purpose. From such features, the design argument then attempts to establish the existence and attributes of an intelligent cause responsible for those features. Just which features signal an intelligent cause, what the nature of that intelligent cause is (e.g., personal agent or telic process) and how convincingly those features establish the existence of an intelligent cause remain subjects for debate and account for the variety of design arguments over the centuries. The design argument is also called the *teleological argument*.

Perhaps the best-known design argument is William Paley's. According to Paley, if one were to find a watch in a field, one could infer by observing the watch's adaptation of parts to telling time that it was designed by an intelligence. So too, according to Paley, the marvelous adaptations of means to ends in organisms ensure that organisms are the products of an intelligence. Paley published this design argument in his 1802 book (which carries a revealing subtitle), *Natural Theology: Evidences of the Existence and Attributes of the Deity, Collected from the Appearances of Nature*. Paley's project was to examine features of the natural world ("appearances of nature") and thereby draw conclusions about the existence and attributes of a designing intelligence responsible for those features (whom Paley identified with the God of Christianity).

Paley's business was natural theology. Intelligent design's business is much more modest: it seeks to identify signs of intelligence to generate scientific insights. Thus, instead of looking to signs of intelligence to obtain theological mileage, as Paley did, intelligent design treats signs of intelligence as strictly part of science. Indeed, within the theory of intelligent design, any appeal to a designer may be viewed as a fruitful device for understanding the world. Construed in this way, intelligent design attaches no significance to questions such as whether a theory of design is in some ultimate sense true, or whether the designer actually exists or what the attributes of that designer are.

Intelligent design is compatible with what philosophers of science call a *constructive empiricist* approach to scientific explanation. Constructive empiricism regards the theoretical entities of science pragmatically rather than realistically. Accordingly, the legitimacy of a scientific entity is tied not to its ultimate reality but to its utility in promoting scientific research and insight. On this view, theoretical entities are constructs with empirical consequences that are scientifically useful to the degree that they adequately account for a range of phenomena.

Scientists in the business of manufacturing theoretical entities like quarks, strings and cold dark matter could therefore view a designer as yet one more theoretical entity in their scientific tool chest. Ludwig Wittgenstein took such an approach. In *Culture and Value* he wrote, "What a Copernicus or a Darwin really achieved was not the discovery of a true theory but of a fertile new point of view." If intelligent design cannot be made into a fertile new point of view that inspires exciting new areas of scientific investigation, then (even if true) it will go nowhere. Yet before being dismissed, intelligent design deserves a fair chance to succeed.

The validity of the design argument, on the other hand, depends not on the fruitfulness of design-theoretic ideas for science but on the metaphysical and theological mileage one can get out of design. A natural theologian might point to nature and say, "Clearly, the designer of this ecosystem prized variety over neatness." A design theorist trying to do actual intelligent design research on that ecosystem might reply, "That's an intriguing theological assertion. Maybe I'll think about that after hours. Right now I'm looking into the sources of information for that variety."

To clarify further the distinction between the design argument and intelligent design, let's review the history of design arguments. First, the design argument needs to be distinguished from a prior metaphysical commitment to design. For instance, in the *Timaeus* Plato (427-347 B.C.) proposed a Demiurge (craftsman) who fashioned the physical world. Plato made this proposal not because the physical world exhibits features that cannot be explained apart from the Demiurge. Plato knew the work of the Greek atomists, who employed no such explanatory device. Rather, within Plato's philosophy, the world of intelligible forms constituted the ultimate reality, of which the physical world was but a dim reflection.

Plato, therefore, posited the Demiurge to explain how the design inherent in the world of forms was mediated to the physical world.

Often the design argument and a metaphysical commitment to design have operated together. This has been especially true in the Christian tradition, in which the design argument is used to establish the existence of an intelligent cause and then a metaphysical commitment to the Christian God identifies that intelligent cause with the Christian God. The design argument and a metaphysical commitment to design have also tended to be conflated within the Christian tradition, so that the design argument often appears to move directly from features of the physical world to the triune God of Christianity.

Full-fledged design arguments have been available since classical times. Both Aristotle's (384-322 B.C.) final causes and the Stoics' seminal reason were types of intelligent causation inferred at least in part from the apparent order and purposiveness of the physical world. Cicero (106-43 B.C.) offers such a design argument in his dialogue *De Natura Deorum*. There he places in the mouth of a Stoic philosopher an argument for the design of the universe based on the machinelike precision and order of the motion of celestial bodies. Cicero's Stoic philosopher attributes this design to a reason that is "transcendent and divine."

Throughout the Christian era, theologians have argued that nature exhibits features which nature itself cannot explain but which instead require an intelligence beyond nature. Church fathers like Minucius Felix (third century A.D.) and Gregory of Nazianzus (c. A.D. 329-389), medieval scholars like Moses Maimonides (1135-1204), and Thomas Aquinas (c. 1225-1274) and commonsense realists like Thomas Reid (1710-1796) and Charles Hodge (1797-1878) all made design arguments, arguing from features of the natural world to an intelligence that transcends nature. Aquinas's fifth proof for the existence of God is perhaps the best known of these.

With the rise of modern science in the seventeenth century, design arguments took a mechanical turn. The mechanical philosophy that was prevalent at the birth of modern science viewed the world as an assemblage of material particles interacting by mechanical forces. Within this view, design was construed as externally imposed form on preexisting inert matter. Paradoxically, the very clockwork universe that early mechanical philosophers like Robert Boyle (1627-1691) used to buttress design in nature was in the end probably more responsible than anything for undermining design in nature. Boyle (in 1686) advocated the mechanical philosophy because he saw it as refuting the immanent teleology of Aristotle and the Stoics, for whom design arose as a natural outworking of natural processes. For Boyle this was idolatry, identifying the source of creation not with God but with nature.

The mechanical philosophy offered a world operating by mechanical principles and processes that could not be confused with God's creative activity and yet allowed such a world to be structured in ways that clearly indicated the divine handiwork and, therefore, design. What's more, the British natural theologians always retained miracles as a mode of divine interaction that could bypass mechanical processes. Over the subsequent centuries, however, what remained was the mechanical philosophy and what fell away was the need to invoke miracles or God as designer. Henceforth, purely mechanical processes could do all the design work for which Aristotle and the Stoics had required an immanent natural teleology and for which Boyle and the British natural theologians required God. In the end, life itself came to be seen not merely as emerging from a mechanical process but as constituting a mechanical process.

The British natural theologians of the seventeenth, eighteenth and nineteenth centuries, starting with Robert Boyle and John Ray (1627-1705) and culminating in the natural theology of William Paley (1743-1805), looked to biological systems for convincing evidence that a designer had acted in the physical world. Accordingly, it was thought incredible that organisms, with their astonishing complexity and superb adaptation of means to ends, could originate strictly through the blind mechanical forces of nature. Paley's *Natural Theology* (1802) is largely a catalog of biological systems he regarded as inexplicable apart from a superintending intelligence. Who was the designer for these British natural theologians? For many it was the traditional Christian God, but for others it was a deistic God, who had created the world but played no ongoing role in governing it.

Criticisms of the design argument have never been in short supply. In classical times, Democritus (c. 460-370 B.C.), Epicurus (c. 342-270 B.C.) and Lucretius (c. 99-55 B.C.) conceived of the natural world as a whirl of particles in collision, which sometimes chanced to form stable configurations exhibiting order and complexity. David Hume (1711-1776) referred to this critique of design as "the Epicurean hypothesis." Modern variants of this critique remain with us in the form of inflationary cosmologies, many-worlds interpretations of quantum mechanics and certain formulations of the anthropic principle (see chapter fifteen).

Though Hume cited the Epicurean hypothesis, he never put great stock in it. In his *Dialogues Concerning Natural Religion* (published posthumously in 1779), Hume argued principally that the design argument fails as an argument from analogy and as an argument from induction (see chapter thirty-two). He also emphasized the problem of imperfect design, or dysteleology. Though widely successful in discrediting the design argument, Hume's critique is no longer as convincing as it used to be.

Hume incorrectly analyzed the logic of the design argument, for the design argument is, properly speaking, neither an argument from analogy nor an argument from induction but an inference to the best explanation. Inference to the best explanation confirms hypotheses according to how well they explain the data under consideration. So staunch a Darwinist as Richard Dawkins (see the beginning of his book *The Blind Watchmaker*) allows that in Hume's day, design was the best explanation for biological complexity.

Whereas Hume attempted a blanket refutation of the design argument, Immanuel Kant (1724-1804) limited the argument's scope. In his *Critique of Pure Reason,* Kant claimed that the most the design argument can establish is "an architect of the world who is constrained by the adaptability of the material in which he works, not a *creator* of the world to whose idea everything is subject." Far from rejecting the design argument, Kant objected to overextending it. For Kant, the design argument legitimately establishes an "architect" (i.e., an intelligent cause whose contrivances are constrained by the materials that make up the world), but it can never establish a creator who originates the very materials that the architect then fashions.

Charles Darwin (1809-1882) delivered the design argument its biggest blow. Darwin was ideally situated historically to do this. His *Origin of Species* (1859) fit perfectly with an emerging positivistic conception of science that was loath to invoke intelligent causes and sought as far as possible to assimilate scientific explanation to unbroken natural law. Hence, even though Darwin's selection mechanism remained much in dispute throughout the second half of the nineteenth century, the mere fact that Darwin had proposed a plausible naturalistic mechanism to account for biological systems was enough to convince the Anglo-American world that some naturalistic story or other had to be true.

Even more than cosmology, biology had, under the influence of British natural theology, become the design argument's most effective stronghold. It was here more than anywhere else that design could assuredly be found. To threaten this stronghold was, therefore, to threaten the legitimacy of the design argument as a reputable intellectual enterprise. Dawkins (in *The Blind Watchmaker*) summed up the matter thus: "Darwin made it possible to be an intellectually fulfilled atheist." God might still exist, but the physical world no longer required God to exist.

Nevertheless, the design argument did not simply wither and die with the rise of Darwinism. Instead, it rooted itself deeper, ramifying into the physical laws that structure the universe. To many scholars of the late-nineteenth and twentieth centuries, thinking of design in terms of biological contrivance was no longer tenable or intellectually satisfying. The focus therefore shifted from finding specific instances of design within the universe to determining whether and in what way the universe as a whole was designed.

The anthropic principle underlies much of the contemporary discussion about the design of the universe. In its original formulation, the anthropic principle (astrophysicist Brandon Carter coined the term in 1970) merely states that the physical laws and fundamental constants that structure the universe must be compatible with human observers. Since human observers exist, the principle is obviously true.

The anthropic principle relates to design because the conditions that need to be satisfied for the universe to permit human observers are so specific that slight variations in these conditions would no longer be compatible with human observers. These conditions are usually cashed out in terms of the laws and fundamental constants of physics. For instance, if the gravitational constant were slightly larger, stars would be too hot and burn too quickly for life to form. On the other hand, if the gravitational constant were slightly smaller, stars would be so cool as to preclude nuclear fusion and with it the production of the heavy elements necessary for life. In either case, human observers would become physically impossible.

The requirement that such precise conditions be satisfied for the existence of human observers seems itself to require explanation and has led to renewed design arguments by both theists and nontheists. (For example, consider the work of Robin Collins, Paul Davies, John Leslie and Richard Swinburne.) They argue that design is the best explanation for the fine-tuning of physical laws and constants.

Nonetheless, using design to explain cosmological fine-tuning (or anthropic coincidences, as they are also called) is controversial. The usual move for refuting such cosmological design arguments is to invoke a selection effect. Accordingly, cosmological fine-tuning is said not to require explanation because without it human observers would not exist to appreciate its absence. Like a lottery in which the winner is pleasantly surprised at being the winner, so human observers are pleasantly surprised to find themselves in a finely tuned universe. No design is required to explain the winning of the lottery, and likewise no design is required to explain human observers residing in a finely tuned universe.

Stated thus, the selection-effect anti-design argument is easily rebutted. What makes chance a viable alternative to design in the lottery analogy is the existence of other lottery players. The reason lottery winners are surprised at their good fortune is that most lottery players are losers. But suppose there was only one lottery player. Suppose that person bought only one lottery ticket and that the probability of that ticket winning was infinitesimally small. What if that person won the lottery? In that case a design inference would follow. That's the situation we're in with the universe. All our empirical evidence points to there just being one lottery player in the grand cosmic lottery—the universe we inhabit.

But is there only one universe? For a selection effect to successfully refute a design argument based on cosmological fine-tuning, one needs a vast ensemble of universes in which most universes are losers in the quest for human observers. But this requires inflating one's ontology, with a consequent blurring of physics and metaphysics, which is itself problematic. The God of most theistic religions is at least thought to be causally active within our universe. Any universes beyond the observable universe, however, are by definition incapable of being observed or otherwise causally impacting us. (For the misbegotten claim that quantum many-worlds causally influence our world through interference effects, see section 2.8 of my book *No Free Lunch*.) Accordingly, these universes are far more an object of faith than any designer purported to influence our universe. Why posit something so fanciful as an ensemble of unobservable and causally disconnected universes, entities that by their very definition can have no observable influence on us? Their principal role seems to consist in short-circuiting the design argument (see chapter fifteen).

One final distinction about the design argument is helpful to keep in mind. Design arguments can focus on whether the universe as a whole is designed. Alternatively, they can focus on whether instances of design have occurred within an already given universe. The universe provides a well-defined causal backdrop. (Physicists these days think of it as a *field* characterized by *field equations*.) Although one can ask whether that causal backdrop is itself designed, one can ask as well whether events and objects occurring within that backdrop are designed. At issue here are two types of design: first, the design of the universe as a whole and, second, instances of design within the universe.

PART 2

DETECTING
DESIGN

8

THE DESIGN INFERENCE

What is the design inference?
How does the design inference differ
from the design argument?

INTELLIGENT DESIGN BEGINS BY RAISING THE FOLLOWING POSSIBILITY: Might there be natural systems that cannot be explained entirely in terms of natural causes and that exhibit features characteristic of intelligence? Such a possibility is certainly legitimate for science to consider. Nonetheless, to evaluate this possibility scientifically, there has to be a reliable way to distinguish between events or objects that result from purely natural causes and events or objects whose emergence additionally requires the help of a designing intelligence. (Note that there is no strict either-or here, as in natural causes *versus* design; at issue is whether natural causes are supplemented or unsupplemented by design.) The whole point of the design inference is to draw such a distinction between natural and intelligent causes.

Specifically, the design inference poses the following question: if an intelligence were involved in the occurrence of some event or in the formation of some object, and if we had no direct evidence of such an intelligence's activity, how could we know that an intelligence was involved at all? The question thus posed is quite general but arises in numerous contexts, including archeology, cryptography, random number generation, SETI (the search for extraterrestrial intelligence) and data falsification in science. I want here to focus on the case of data falsification involving Hendrik Schön. This case points up the legitimacy of the design inference and ties it to a matter of real urgency facing the scientific community.

On May 23, 2002, the *New York Times* reported on the work of "J. Hendrik Schön, 31, a Bell Labs physicist in Murray Hill, N.J., who has produced an extraordinary body of work in the last two and a half years, including seven articles each in *Science* and *Nature*, two of the most prestigious journals." Despite such impeccable credentials, Schön's career was on the line. Why? According to the *New York Times*, Schön published "graphs that were nearly identical even though they appeared in different scientific papers and represented data from different devices. In some graphs, even the tiny squiggles that should arise from purely random fluctuations matched exactly." As a consequence, Bell Labs appointed an independent panel to determine whether Schön "improperly manipulat[ed] data in research papers published in prestigious scientific journals." In September 2002 the panel concluded that Schön was indeed guilty of data falsification. Bell Labs then fired Schön.

The theoretical issues raised in this case are precisely those addressed by the design inference. To determine whether Schön cooked the numbers, the panel had to note two things: that the first published graph provided an independently given pattern or specification for the second; and that the match between the two graphs in Schön's articles was highly improbable (or "complex," in the sense described in chapter one) under the supposition that the graphs resulted from random fluctuations. The randomness here is well understood and would have operated had Schön performed the experiments as he claimed. Consequently, the panel set aside searching for an unknown material mechanism or natural process to explain how the graphs from independent experiments on independent devices could have exhibited the same pattern of random fluctuations. Instead, the panel rationally concluded data manipulation and design. In other words, it reached that conclusion by identifying a highly improbable, independently given pattern, or what we call *specified complexity.*

There's no way to get around specified complexity when inferring design. Design inferences arise in cases where the evidence is circumstantial and thus where we lack direct evidence of a designing intelligence. Where direct evidence is missing, there is no problem explaining an event as the result of chance, even if it is highly improbable or complex. Highly improbable unspecified events, after all, happen by chance all the time. Just get out a coin and flip it a thousand times. The precise sequence of heads and tails you observe is unimaginably improbable, less than one chance in 10^{300}. But that sequence is also unspecified. Invoking chance to explain such an event only becomes a problem when it is not only highly improbable or complex, but also matches an independently given pattern or specification.

In the case of Schön's graphs, under the relevant chance hypotheses characterizing the random fluctuations, the match between graphs needed to be highly improbable to cast suspicion on Schön. (If the graphs were merely two-bar histograms with only a few possible gradations in height, then a match between the graphs would be reasonably probable and no one would ever have questioned Schön's integrity.) The panel determined that indeed it was highly improbable that the graphs matched so precisely. Improbability, however, wasn't enough. The random fluctuations of each graph taken individually are indeed highly improbable. But it was the match between the graphs that raised suspicion. That match renders one graph a specification for the other so that in the presence of improbability, a design inference is warranted.

By itself the design inference does not implicate any particular intelligence. A design inference would show that the data in Schön's papers were improperly manipulated. It could not, however, show that Schön was the culprit. To identify the actual culprit requires a more detailed causal analysis—an analysis that in the Schön case was conducted by Bell Labs' independent panel. On the basis of that analysis, the panel concluded that Schön was indeed the culprit. Indeed, not only was he the first author on the articles in question, but he alone among his coauthors had access to the devices that produced the disconcertingly coincident experimental outcomes. What's more, all the experimental protocols that were Schön's responsibility to maintain mysteriously disappeared when the panel wanted to review them.

The design inference drawn by the independent panel in the Schön case illustrates the difference between the design inference and the design argument. The design argument is at its heart a philosophical and theological argument. It attempts to establish the existence and attributes of an intelligent cause behind the world based on certain features of the world. By contrast, the design inference is a generic argument for identifying the effects of intelligence regardless of the intelligence's particular characteristics and regardless of where, when, how or why the intelligence acts. (The intelligence can be animal, human, extraterrestrial, singular, plural, immanent or transcendent.) The design inference looks to one feature in particular—specified complexity—and uses it as the basis for inferring intelligence. Thus, when an event, object or structure in the world exhibits specified complexity, one infers that an intelligence was responsible for it. In other words, one draws a design inference.

9

CHANCE AND NECESSITY

*How does the scientific community
conceive of natural causes, and why aren't
intelligent causes among them?*

THE SCIENTIFIC COMMUNITY UNDERSTANDS natural causes in terms of chance, necessity and their combination. The biologist Jacques Monod even wrote a book to emphasize that point: *Chance and Necessity*. Why does the scientific community limit natural causes to chance and necessity, thereby excluding design? For many in the natural sciences, design, as the action of an intelligent agent, is not a fundamental creative force in nature. Rather, blind natural forces, characterized by chance and necessity, are thought sufficient to do all of nature's creating.

Three objects help illustrate what's at stake with chance and necessity: a calculator, a roulette wheel and an agitator. A calculator operates by necessity. Punch in 2 + 2 and the only possible output is 4. To say that something is necessary is to say that it has to happen and that it can happen in one and only one way. It's necessary that water cooled below zero degrees Celsius freezes. It's necessary that massive objects in a gravitational field are attracted by gravitational forces. It's necessary that two plus two equals four.

The opposite of necessity is contingency. To say that something is contingent is to say that it can happen in more than one way. When you spin a roulette wheel, for instance, the ball can land in any one of thirty-eight slots (numbered one through thirty-six as well as zero and double-zero). The precise place where the ball lands is therefore contingent. Contingency presupposes a range of possibilities. To get a handle on those possibilities, scientists typically assign them probabilities. With the roulette wheel, each slot has the same probability of the ball landing in it. When all possibilities have the same probability, we call this *pure chance*, or *randomness*. Pure chance or randomness therefore characterizes the behavior of the roulette wheel.

Pure chance or randomness is the exception rather than the rule. Most of the time when natural causes operate, chance and necessity act together rather than in isolation. Probabilities still apply, but things are not as straightforward as when all possibilities have the same probability. Consider, therefore, the third of our objects—an agitator. An agitator is a container into which one places rocks and then shakes them. This shaking (or agitating) constitutes a purely chance-driven process that, apart from any further constraints, would yield a totally disordered arrangement of rocks.

Here on earth, however, there is a further constraint, namely, gravity. Agitating the container in the presence of a downward-directed gravitational force ensures that the largest rocks will rise to the top and that the smallest will proceed to the bottom. Precisely where at a given level a rock of the same size as others lands will still be random. Thus randomness or pure chance still operates horizontally. But vertically there is a clear order imposed on rocks agitated in the presence of gravity.

The purpose of an agitator is to sift rocks according to size. To that end chance and necessity must act together: necessity here takes the form of gravity, and chance takes the form of random shaking. By randomly shaking a container filled with rocks in a gravitational field, chance and necessity produce an ordered arrangement of rocks with the smallest rocks on the bottom, medium-sized rocks in the middle and the largest rocks at the top.

The agitator example is relevant to Darwinism. Darwin's main claim to fame was to propose a theory for how the joint action of chance and necessity might agitate matter toward biological complexity. Darwin opened his *Origin of Species* by describing animal breeding experiments. The offspring of animals, though similar to their parents, nonetheless differ from them. These differences for all we know are controlled by chance. (Darwin referred to these differences as variations; biologists now think of them as arising from mutations in DNA, which is more evidently chance driven. Darwin did not speculate about how variations arose.)

Now, if animal breeders flipped a coin whenever they decided which of their animals could reproduce, then any chance difference between parents and offspring would remain unconstrained. But if, instead, animal breeders monitored their animals and permitted only those to reproduce that, albeit by chance, exhibited certain desirable characteristics, then over time those desirable characteristics would become intensified and entrenched in the population. Thus chance, in the form of variation between parent and offspring, would be shaped or constrained by the animal breeder to yield those characteristics that the animal breeder prefers. Note that the animal breeder is limited to those characteristics that chance produces. A breeder of furry animals, for instance, will not be able to obtain smooth-skinned animals if all the animals bred stay unvaryingly furry.

Human animal breeders are of course absent from most of the history of life. What, then, constrains the reproduction of living things in nature? Clearly, nature itself. Nature, as it were, selects those organisms that will reproduce and eliminates those that will not. Since the reproduction of organisms would, apart from constraints, proceed exponentially, and since nature clearly does not have the resources to accommodate such exponential growth, only a small proportion of organisms will in any generation successfully reproduce. Those whose characteristics best aid reproduction will therefore be selected to leave offspring, whereas the rest will die without leaving them. Thus, according to Darwin, nature itself constitutes the supreme animal breeder that shapes the path of life. In particular, necessity, in the form of natural selection, and chance, in the form of random variation, are said to account for all of biological complexity and diversity.

Where does design or intelligent causation fit within this dialectic between chance and necessity? It doesn't. At best design becomes a byproduct of chance and necessity. If, for instance, the Darwinian mechanism of random variation and natural selection accounts for the emergence of human beings, then human intelligence (with all its design capabilities) is merely a complex behavioral capacity that sits atop blind material processes.

To be sure, it is a logical possibility that purpose, intelligence and design arise purely through chance and necessity. Accordingly, intelligence might be merely a survival tool given to us by a Darwinian evolutionary process that places a premium on survival and reproduction and that is itself not intelligently guided but driven solely by chance and necessity. The basic creative forces of nature would thus be devoid of intelligence. Yet even though this is a logical possibility, it clearly is not the only possibility. Another logical possibility is that purpose, intelligence and design are fundamental features of reality and are not reducible to chance and necessity. Which is the right one? Instead of prejudging the answer as Darwinism does, the design inference provides a logical framework for assessing which of these possibilities holds. Darwinism rules out design from biology. The design inference, by contrast, neither rules it out nor requires it. Rather, it allows the evidence of biology to decide it.

10

SPECIFIED COMPLEXITY

What is specified complexity, and how does one determine whether something exhibits specified complexity?

THE TERM *SPECIFIED COMPLEXITY* is about thirty years old. To my knowledge, origin-of-life researcher Leslie Orgel was the first to use it. The term appeared in his 1973 book *The Origins of Life,* where he wrote, "Living organisms are distinguished by their specified complexity. Crystals such as granite fail to qualify as living because they lack complexity; mixtures of random polymers fail to qualify because they lack specificity." More recently, in his 1999 book *The Fifth Miracle,* Paul Davies identified specified complexity as the key to resolving the problem of life's origin:

> Living organisms are mysterious not for their complexity *per se,* but for their tightly specified complexity. To comprehend fully how life arose from nonlife, we need to know not only how biological information was concentrated, but also how biologically useful information came to be specified.

Neither Orgel nor Davies, however, provided a precise analytic account of specified complexity. I provide such an account in *The Design Inference* (1998) and its sequel *No Free Lunch* (2002). Here I will merely sketch that account of specified complexity. Orgel and Davies used the term *specified complexity* loosely. In my own research I've formalized it as a statistical criterion for identifying the effects of intelligence. Specified complexity, as I develop it, incorporates five main ingredients:

- a probabilistic version of complexity applicable to events
- conditionally independent patterns
- probabilistic resources, which come in two forms: replicational and specificational

- a specificational version of complexity applicable to patterns
- a universal probability bound

Let's consider these briefly.

Probabilistic complexity. Probability can be viewed as a form of complexity. To see this, consider a combination lock. The more possible combinations of the lock, the more complex the mechanism and, correspondingly, the more improbable that the mechanism can be opened by chance. For instance, a combination lock whose dial is numbered from zero to thirty-nine and that must be turned in three alternating directions will have 64,000 (i.e., 40 x 40 x 40) possible combinations. This number gives a measure of complexity for the combination lock but also corresponds to a 1 in 64,000 probability of the lock being opened by chance (assuming zero prior knowledge of the lock combination). A more complicated combination lock whose dial is numbered from zero to ninety-nine and which must be turned in five alternating directions will have 10,000,000,000 (i.e., 100 x 100 x 100 x 100 x 100) possible combinations and thus a 1 in 10,000,000,000 probability of being opened by chance. Complexity and probability therefore vary inversely: the greater the complexity, the smaller the probability. The *complexity* in *specified complexity* refers to improbability.

Conditionally independent patterns. The patterns that in the presence of complexity or improbability implicate a designing intelligence must be independent of the event whose design is in question. Crucial here is that patterns not be artificially imposed on events after the fact. For instance, if an archer shoots arrows at a wall and we then paint targets around the arrows so that they stick squarely in the bull's-eyes, we impose a pattern after the fact. Any such pattern is not independent of the arrow's trajectory. On the other hand, if the targets are set up in advance ("specified") and then the archer hits them accurately, we know it was not by chance but rather by design (provided, of course, that hitting the targets is sufficiently improbable). The way to characterize this independence of patterns is via the probabilistic notion of conditional independence. A pattern is conditionally independent of an event if adding our knowledge of the pattern to a chance hypothesis does not alter the event's probability under that hypothesis. The *specified* in *specified complexity* refers to such conditionally independent patterns. These are the specifications.

Probabilistic resources. Probabilistic resources refer to the number of opportunities for an event to occur or be specified. A seemingly improbable event can become quite probable once enough probabilistic resources are factored in. On the other hand, such an event may remain improbable even after all the available probabilistic resources have been factored in. Think of trying to deal yourself a royal flush. Depending on how many hands you can deal, that outcome, which by itself is quite improbable, may remain improbable or become quite probable. If you can only deal yourself a few dozen hands, then in all likelihood you won't see a royal flush. But if you can deal yourself millions of hands, then you'll be quite likely to see it.

Probabilistic resources come in two varieties: replicational and specificational. Replicational resources refer to the number of opportunities for an event to occur. Specificational resources refer to the number of opportunities to specify an event. To see what's at stake with these two types of probabilistic resources, imagine a large wall with N identically sized, non-overlapping targets painted on it and M arrows in your quiver. Let's say that your probability of hitting any one of these targets, taken individually, with a single arrow by chance is p. Then the probability of hitting any one of these N targets, taken collectively, with a single arrow by chance is bounded by Np (i.e., N and p multiplied); and the probability of hitting any of these N targets with at least one of your M arrows by chance is bounded by MNp (i.e., M and N and p multiplied). In this case, the number of replicational resources corresponds to M (the number of arrows in your quiver), the number of specificational resources corresponds to N (the number of targets on the wall), and the total number probabilistic resources corresponds to the product MN. For a specified event of probability p to be reasonably attributed to chance, the number MNp must not be too small.

Specificational complexity. Because they are patterns, specifications exhibit varying degrees of complexity. A specification's degree of complexity determines how many specificational resources must be factored in when gauging the level of improbability needed to preclude chance (see the previous point). The more complex the pattern, the more specificational resources must be factored in. The details are technical and involve a generalization of what mathematicians call *Kolmogorov complexity.* Nevertheless, the basic intuition is straightforward. Low specificational complexity is important in detecting design because it ensures that an event whose design is in question was not simply described after the fact and then dressed up as though it could have been described before the fact.

To see what's at stake, consider the following two sequences of ten coin tosses: HHHHHHHHHH and HHTHTTTHTH. Which of these would you be more inclined to attribute to chance? Both sequences have the same

probability, approximately 1 in 1,000. Nevertheless, the pattern that specifies the first sequence is much simpler than the second. For the first sequence the pattern can be specified with the simple statement "ten heads
in a row." For the second sequence, on the other hand, specifying the pattern requires a considerably longer statement, for instance, "two heads,
then a tail, then a head, then three tails, then heads followed by tails and
heads." Think of specificational complexity (not to be confused with specified complexity) as minimum description length. (For more on this, see
<www.mdl-research.org>.)

For something to exhibit specified complexity it must have low specificational complexity (as with the sequence HHHHHHHHHH, consisting
of ten heads in a row) but high probabilistic complexity (i.e., its probability
must be small). It's this combination of low specificational complexity (a
pattern easy to describe in relatively short order) and high probabilistic
complexity (something highly unlikely) that makes specified complexity
such an effective triangulator of intelligence. But specified complexity's
significance doesn't end there.

Besides its crucial place in the design inference, specified complexity
has also been implicit in much of the self-organizational literature, a field
that studies how complex systems emerge from the structure and dynamics of their parts. Because specified complexity balances low specificational complexity with high probabilistic complexity, specified complexity
sits at that boundary between order and chaos commonly referred to as the
"edge of chaos." The problem with pure order (low specificational complexity) is that it is predictable and thus largely uninteresting. An example
here is a crystal that keeps repeating the same simple pattern over and
over. The problem with pure chaos (high probabilistic complexity) is that
it is so disordered that it is also uninteresting. (No meaningful patterns
emerge from pure chaos. An example here is the debris strewn by a tor-
nado or avalanche.) Rather, it's at the edge of chaos, neatly ensconced between order and chaos, that interesting things happen. That's where specified complexity sits.

Universal probability bound. In the observable universe, probabilistic
resources come in limited supplies. Scientists estimate that within the
known physical universe, there are around 10^{80} elementary particles.
Moreover, the properties of matter are such that transitions from one physical state to another cannot occur at a rate faster than 10^{45} times per second.
This frequency corresponds to the Planck time, which constitutes the
smallest physically meaningful unit of time. Finally, the universe itself is

about a billion times younger than 10^{25} seconds (assuming the universe is between 10 and 20 billion years old). If we now assume that any specification of an event within the known physical universe requires at least one elementary particle to specify it and that such specifications cannot be generated any faster than the Planck time, then these cosmological constraints imply that the total number of specified events throughout cosmic history cannot exceed $10^{80} \times 10^{45} \times 10^{25} = 10^{150}$. Thus, any specified event of probability less than 1 in 10^{150} will remain improbable even after all conceivable probabilistic resources from the observable universe have been factored in. A probability of 1 in 10^{150} is therefore a *universal probability bound*. (For the details justifying this universal probability bound, see my 1998 book *The Design Inference*.) A universal probability bound is impervious to all available probabilistic resources that may be brought against it. Indeed, all the probabilistic resources in the known physical world cannot conspire to render remotely probable an event whose probability is less than this universal probability bound.

The universal probability bound of 1 in 10^{150} is the most conservative in the literature. The French mathematician Emile Borel proposed 1 in 10^{50} as a universal probability bound below which chance could definitively be precluded. (That is, any specified event as improbable as this could never be attributed to chance.) Cryptographers assess the security of cryptosystems in terms of brute force attacks that employ as many probabilistic resources as are available in the universe to break a cryptosystem by chance. In its report on the role of cryptography in securing the information society, the National Research Council set 1 in 10^{94} as its universal probability bound to ensure the security of cryptosystems against chance-based attacks. (See Kenneth Dam and Herbert Lin's 1996 book *Cryptography's Role in Securing the Information Society*.) Computer scientist Seth Lloyd sets 10^{120} as the maximum number of bit-operations that the universe could have performed throughout its entire history (*Physical Review Letters*, June 10, 2002). That number corresponds to a universal probability bound of 1 in 10^{120}. Stuart Kauffman in his most recent book, *Investigations* (2000), comes up with similar numbers.

For something to exhibit specified complexity therefore means that it matches a conditionally independent pattern (i.e., specification) of low specificational complexity, but where the event corresponding to that pattern has a probability less than the universal probability bound and therefore high probabilistic complexity. Specified complexity is a widely used criterion for detecting design. For instance, when researchers in the search for extraterrestrial intelligence (SETI) look for signs of intelligence from outer space, they are looking for specified complexity. (Recall the movie *Contact* in which SETI detects an intelligent signal pattern when a long sequence of prime numbers comes in from outer space. Such a sequence exhibits specified complexity.)

11

THE EXPLANATORY FILTER

*How does specified complexity function as
a criterion for detecting design?*

WHEN TRYING TO EXPLAIN SOMETHING, we employ three broad modes of explanation: *necessity, chance* and *design*. As a criterion for detecting design, specified complexity enables us to decide which of these modes of explanation apply. It does that by answering three questions about the thing we are trying to explain: Is it contingent? Is it complex? Is it specified? By arranging these questions sequentially as decision nodes in a flowchart, we can represent specified complexity as a criterion for detecting design. This flowchart is now widely known as the Explanatory Filter (see figure 1).

The Explanatory Filter works by feeding in events at the "start" node and then sending them past the decision nodes. Suppose, for instance, we want to explain why a certain well-constructed safe with a combination lock happens to be open. How would we explain the opening of the safe? Let us suppose that the safe's combination lock is marked with a hundred numbers ranging from 00 to 99 and that five turns in alternating directions are required to open the lock. We assume that one and only one sequence of alternating turns is capable of opening the lock (e.g., 34-98-25-09-71). There are thus 10 billion possible combinations, of which precisely one opens the lock.

Let's now feed the opening of the bank's safe into the Explanatory Filter. How does this event fare at the first decision node? Since no regularity or law of nature requires that the combination lock turn to the combination that opens it, the opening of the bank's safe is contingent. This event therefore moves past the first decision node. How does it fare at the second decision node? With 10 billion possibilities, only one of which opens the safe, random twirling of the combination lock's dial is exceedingly unlikely to open the lock. For practical purposes, the opening of the safe is therefore complex or improbable in the required sense. (Strictly speaking, the probability would have to be less than the universal probability bound, but for

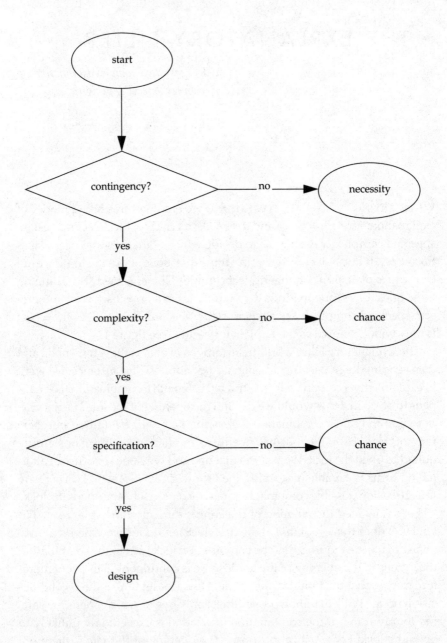

Figure 1. The Explanatory Filter

practical purposes this is enough since anyone dialing the combination has very few opportunities to get it right, as explained in the previous chapter.)

The opening of the bank's safe therefore lands at the third decision node. Here the crucial question is whether this event is also specified. If it were only complex but not specified, then the opening of the bank's safe could legitimately be attributed to chance. (Chance readily accounts for sheer complexity apart from specification.) But of course this event is specified. Indeed, the very construction of the lock's tumblers specifies which one of the 10 billion combinations opens the lock. This event is therefore both complex and specified. As such, it exhibits specified complexity and passes to the terminal "design" node. Thus we arrive formally at what any sane bank worker would instantly recognize: somebody knew, and chose to dial, the correct combination.

The Explanatory Filter has come under considerable criticism both in print and on the Internet. I want therefore to respond here briefly to the main objections. One concern, expressed by people who've only heard about the filter second-hand, is that it assigns merely improbable events to design. But that clearly is not the case since, in addition to identifying complexity or improbability, the filter needs to identify a specification before attributing design. Another concern is that the filter will assign to design regular geometric objects like the star-shaped ice crystals that form on a cold window. This criticism fails because such shapes form as a matter of physical necessity simply in virtue of the properties of water. (The filter will therefore assign the crystals to necessity and not to design.) Similar considerations apply to self-organizing systems generally.

According to Internet critic Gert Korthof (<home.planet.nl/~gkorthof/kortho44.htm>), the filter mistakenly attributes design to certain regular arithmetic sequences that arise in the growth of biological systems—sequences that instead ought to be attributed to natural necessities. For instance, Fibonacci sequences (for which each number is the sum of the two previous numbers) characterize the arrangement of leaves on the stems of certain plants. This simple observation, according to Korthof, implicates the Explanatory Filter in a contradiction. On the one hand, he contends, the filter attributes such Fibonacci sequences to design (just as the sequence of prime numbers that in the movie *Contact* was attributed to design). On the other hand, the filter attributes such sequences to necessity since, as Korthof claims, the arrangement of leaves on the plant's stem derives from a "perfectly natural process."

The contradiction here is only apparent. The fault is not with the Explanatory Filter but with an equivocation involving the term *natural*. In what sense do Fibonacci sequences derive from a "perfectly natural process"? Is the operation of that process natural? Or is the origin of that process natural? Just because the operation of something is natural does not mean that its origin is. This is a widely neglected point. In fact, Korthof's example speaks only to the operation of biological systems being natural, not to their origin. Yet the origin, and therefore design, of biological systems that display Fibonacci behavior is itself in question. Korthof's example is logically equivalent to a computer programmed to output Fibonacci sequences. Once suitably programmed, the computer operates by necessity. Consequently, its outputs, when fed into the filter, will land at the necessity node of the filter. The computer-generated Fibonacci sequences derive, as Korthof might put it, from a "perfectly natural process." But whence the computer that runs the program? And whence the program? All the computer hardware and software in our ordinary experience is properly referred not to necessity but to design.

Korthof's criticism points up that the conclusions we derive from the Explanatory Filter depend crucially on the events we feed into it. To examine any putative design behind biologically produced Fibonacci sequences, the event of interest is not the programmed operation of biological systems that outputs Fibonacci sequences (which, treated as such, would land at the necessity node of the Explanatory Filter). Rather, the event of interest is the structuring event that arranged biological systems so that they could output Fibonacci sequences in the first place. To see that this event lands at the design node of the Explanatory Filter is straightforward. Granted, the "biological software" that outputs Fibonacci sequences is probably quite simple and might even be due to purely natural forces like selective pressure. (It's a stretch, but let's grant it for the sake of argument.) Nevertheless, the simplest "biological hardware" that runs that software is a functioning cell. And the simplest functioning cell is staggeringly complex, exhibiting layer upon layer of specified complexity and therefore design.

Korthof's failure to distinguish between the lawlike or "natural" operation of a thing and its designed origin is widespread. Unfortunately, that confusion shows no signs of abating. Michael Polanyi addressed the confusion as far back as the 1960s in his distinction between the mechanistic operation of organisms and their machinelike features (see chapter twenty). Let me therefore spell out the distinction with a particularly simple example. Imagine an embossed sign that reads *Eat at Frank's* falls over in a snowstorm and leaves the mirror image of *Eat at Frank's* embedded in the snow. Granted, the sign fell over as a result of undirected natural forces, and on that basis the impression the sign left in the snow would not be attributed to design by the filter. Nonetheless, there is a relevant event whose design needs to be assessed, namely, the structuring of the embossed image (whether in the snow or on the sign). This event must be referred back to the activity of the sign's maker, and the Explanatory Filter properly ascribes it to design. Natural forces can serve as conduits of design. As a result, a simple inspection of those natural forces may turn up no evidence of design. Often one must look deeper. To use the Explanatory Filter to identify design requires inputting the right events, objects and structures into the filter. Just because one input does not turn up design does not mean that another, more appropriately chosen input, will fail.

In an article titled "The Advantages of Theft over Toil: The Design Inference and Arguing from Ignorance" for the journal *Biology and Philosophy* (2001), John Wilkins and Wesley Elsberry argue that the filter is not a reliable indicator of design. Central to their argument is that if we fail to characterize the full range of natural necessities and chance processes that might have been operating to account for a phenomenon, we may omit an undirected natural cause that renders the phenomenon likely and thereby adequately accounts for it apart from design. Thus, with the combination lock example given above, they consider a poorly constructed lock for which the probability of opening it by chance is much larger than one in 10 billion. Granted, this could happen. But it could also happen that the lock requires dialing the right combination so precisely that the chance of opening it by chance is in fact far smaller than one in 10 billion. Further investigation of the lock could therefore upset or reinforce a design inference.

The prospect that further knowledge will upset a design inference poses a risk for the Explanatory Filter. But it is a risk endemic to all of scientific inquiry. Indeed, it merely restates the problem of induction, namely, that we may be wrong about the regularities (be they probabilistic or necessitarian) which operated in the past and apply in the present. Wilkins and Elsberry act as if no amount of investigation into a phenomenon is enough to reasonably rule out natural necessities and chance processes as its cause. Yet if design in nature is real, their recommendation ensures we'll never see it (see chapters twenty-six and thirty-two).

Contrary to Wilkins and Elsberry, the risk that further knowledge may upset a design inference has nothing to do with the filter's reliability. The filter's reliability refers to its accuracy in detecting design provided we have accurately assessed the probabilities in question (see chapter twelve). Wilkins and Elsberry purport to criticize the filter's reliability but are in fact criticizing its applicability (see chapter fourteen). They're like someone who dismisses a calculator as unreliable after a friend, seeking to know what nine times nine is, gets the wrong answer by accidentally punching "6 x 6." If that person were dead set on dismissing the calculator's usefulness but were pressed to admit that the friend had made the error, then the calculator hater might insist that nobody can be trusted to use the calculator accurately. That's essentially what Wilkins and Elsberry have done with the Explanatory Filter.

To so refuse the Explanatory Filter's applicability irrationally privileges undirected natural causes and renders them immune to disconfirmation. Science is supposed to be a risky enterprise. We go to nature to discover its secrets because we do not know what those secrets are till we look. It follows that what nature reveals to us can be unexpected and even disconcerting. (Witness the difficulties physicists faced making sense of quantum mechanics in the 1920s and 1930s.) Yet when it comes to design, Wilkins and Elsberry want a risk-free science. They want their science safely cosseted within a naturalistic cocoon that excludes any place for design in the natural sciences. But such a risk-free science is no science at all. It knows the truth without looking. So when evidence comes that challenges it, it arbitrarily rules that evidence inadmissible.

Next, critics object that in distinguishing chance and necessity, the filter fails to account for the joint action of chance and necessity, especially as these play out in the Darwinian mechanism of natural selection (the necessity component) and random variation (the chance component). In particular, the Darwinian mechanism is supposed to deliver all the biological complexity that the filter attributes to design. If correct, this objection would overthrow the Explanatory Filter. But it is not correct. I approach chance and necessity as a probabilist for whom necessity is a special case of chance in which probabilities collapse to zero and one. (Think of a double-headed coin: what is the probability that it will land heads? What is the probability that it will land tails?) Chance as I characterize it thus includes necessity, chance (as it is ordinarily used) and the combination of these. The filter could therefore be compressed by assimilating the necessity node into the chance nodes, though at the expense of making the filter less user-friendly. At any rate, the filter is robust and fully applicable to evaluating the claims of Darwinism.

Finally, Michael Ruse charges that the Explanatory Filter makes necessity, chance and design mutually exclusive and exhaustive. Citing Ronald Fisher, Ruse argues that all three need to be "run together." Thus, in *Can a Darwinian Be a Christian?* he writes,

> [Fisher] believed that mutations come individually by chance, but that collectively they are governed by laws (and undoubtedly are governed by the laws of physics and chemistry in their production) and thus can provide the grist for selection (law) which produces order out of disorder (chance). He cast the whole picture within the confines of his "fundamental theory of natural selection," which essentially says that evolution progresses upwards, thus countering the degenerative processes of the Second Law of Thermodynamics. And then, for good measure, he argued that everything was planned by his Anglican God!

This, then, is how Fisher ran together chance (conceived as mutations and disorder generally), necessity (conceived as physical laws and natural selection) and design (conceived as the planning of an Anglican God). Is such an explanatory stew compatible with the Explanatory Filter? In fact it is.

Ruse is wrong that the Explanatory Filter separates necessity, chance and design into mutually exclusive and exhaustive categories. The filter models our ordinary practice of ascribing these modes of explanation. Of course all three can be run together. But typically one of these modes of explanation predominates. Is the rusted old automobile in your driveway designed? The rust and the automobile's beaten appearance are due to chance and necessity (weathering, gravity and a host of other unguided natural forces). But your automobile also exhibits design, which typically is the point of interest. What's more, by focusing on suitable aspects of the automobile, the filter detects that design. Ultimately, what enables the filter to detect design is specified complexity. The Explanatory Filter provides a user-friendly way to establish specified complexity. For that reason, the only way to refute the Explanatory Filter is to show that specified complexity is an inadequate criterion for detecting design. Let's now turn to that worry.

12

RELIABILITY OF THE CRITERION

Is specified complexity a reliable criterion for detecting design?

SPECIFIED COMPLEXITY, AS ENCAPSULATED IN THE Explanatory Filter (see chapter eleven), is a criterion for detecting design. I refer to this criterion as the complexity-specification criterion. In general, criteria attempt to classify individuals with respect to some target group. The target group for the complexity-specification criterion comprises all things intelligently caused. How accurate is this criterion at correctly assigning things to this target group and correctly omitting things from it?

The things we are trying to explain have causal histories. In some of those histories, intelligent causation is indispensable whereas in others it is dispensable. An inkblot can be explained without appealing to intelligent causation; ink arranged to form meaningful text cannot. When the complexity-specification criterion assigns something to the target group, can we be confident that it actually is intelligently caused? If not, we have a problem with false positives. On the other hand, when this criterion fails to assign something to the target group, can we be confident that it is not intelligently caused? If not, we have a problem with false negatives.

Consider first the problem of false negatives. When the complexity-specification criterion fails to detect design in a thing, can we be sure that no intelligent cause played a role in its formation? No, we cannot. To determine that something is not designed, this criterion is not reliable. False negatives are a problem for it. This problem of false negatives, however, is endemic to design detection in general. One difficulty is that intelligent causes can mimic undirected natural causes, thereby rendering their actions indistinguishable from such unintelligent causes. A bottle of ink happens to fall off a desk and accidentally spills onto a sheet of paper. Alternatively, a human agent deliberately takes a bottle of ink and pours it over a sheet of paper. The resulting inkblots may look identical, with neither providing evidence of design. Yet one inkblot actually is due to design whereas the other is due merely to natural causes.

Another difficulty is that detecting intelligent causes requires background knowledge. It takes an intelligent cause to recognize an intelligent cause. But if we do not know enough, we will miss it. Consider a spy listening in on a communication channel whose messages are encrypted. Unless the spy knows how to break the cryptosystem used by the parties on whom she is eavesdropping (i.e., unless she knows the cryptographic key), any messages traversing the communication channel will be unintelligible and appear random rather than designed.

The problem of false negatives therefore arises either when an intelligent agent has acted (whether consciously or unconsciously) to conceal its actions or when an intelligent agent, in trying to detect design, lacks knowledge essential for detecting it. Thus, the complexity-specification criterion does have its limits. It's just that those limits do nothing to save the Darwinist from the crucial work the criterion can do. This criterion is fully capable of detecting intelligent causes intent on making their presence evident—and even many that aren't. Masters of stealth intent on concealing their actions may successfully evade the criterion. But masters of self-promotion bank on the complexity-specification criterion to make sure their intellectual property gets properly attributed. Indeed, intellectual property law (like patent and copyright protection) would be impossible without this criterion.

And that brings us to the problem of false positives. Even though specified complexity is not a reliable criterion for *eliminating* design, it is a reliable criterion for *detecting* design. The complexity-specification criterion is a net. Things that are designed will occasionally slip past the net. We would prefer that the net catch more than it does, omitting nothing due to design. But given the ability of design to mimic unintelligent causes and the possibility of our passing over in ignorance things that are designed, this problem cannot be remedied. Nevertheless, we want to be very sure that whatever the net does catch includes only what we intend it to catch, namely, things that are designed. If that is the case, we can have confidence that whatever the complexity-specification criterion attributes to design is indeed designed.

How can we see that specified complexity is a reliable criterion for detecting design? In other words, how can we see that the complexity-specification criterion successfully avoids false positives? The justification for this claim is a straightforward inductive generalization: in every instance where specified complexity is present and where the underlying causal story is known (i.e., where we are not just dealing with circumstantial evidence, but where, as it were, the video camera is running and any putative designer would be caught red-handed), it turns out design is present as well. This is true even where the person running the filter isn't privy to the firsthand information. That's a bold and fundamental claim, so I'll restate it: *Where direct, empirical corroboration is possible, design actually is present whenever specified complexity is present.*

Although this justification of the complexity-specification criterion's reliability at detecting design may seem a bit too easy, it really isn't. If something genuinely exhibits specified complexity, then one can't explain it in terms of all material mechanisms (not only those that are known but all of them, thanks to the universal probability bound of 1 in 10^{150}; see chapter ten). Indeed, to attribute specified complexity to something is to say that the specification to which it conforms corresponds to an event that is vastly improbable with respect to all material mechanisms that might give rise to the event. So take your pick—treat the item in question as inexplicable in terms of all material mechanisms, or treat it as designed. But since design is uniformly associated with specified complexity when the underlying causal story is known, induction counsels attributing design in cases where the underlying causal story is not known.

For specified complexity to detect design, it is not enough that the probability be small with respect to some arbitrarily chosen probability distribution. For instance, if one encountered a forest with trees where the moss consistently stood on the north sides of the trunk, it wouldn't be enough to say, "What are the odds that the moss would just happen to be on the north side of every one of these 10,000 trees? It's virtually impossible. Somebody must have come along and scraped off all the moss on the south side of all 10,000 trunks." Our wannabe forest ranger here has overlooked another probability distribution, which takes into account that the moss only gets enough shade to thrive on the north side of the trunks.

Thus, for specified complexity to detect design, the probability of the thing in question must be small with respect to every probability distribution that might characterize it. Where that is the case, a design inference follows. The use of chance here is very broad and includes anything that can be captured mathematically by a stochastic process. (Stochastic processes constitute the most general mathematical model for describing the interplay of chance and necessity over time.) It thus includes deterministic processes whose probabilities all collapse to zero and one (such as necessity, regularities and natural laws). It also includes nondeterministic processes, like evolutionary processes that combine random variation and natural selection. Indeed, chance so construed characterizes all undirected natural processes. In eliminating chance, specified complexity therefore sweeps the field of all processes that could preclude design. The only reasonable possibility left, in that case, would be design.

But that still leaves the problem of dispensing with the probability distributions induced by material mechanisms. Can this be done with confidence? If the probability distributions in question are those induced by known material mechanisms operating in known ways, then specified complexity can and indeed must dispense with them. (Specified complexity would be immediately overturned if some probability distribution induced by known mechanisms operating in known ways rendered the thing we're trying to explain reasonably probable. Recall the tree moss example.) But since specified complexity also requires eliminating all probability distributions induced by any material mechanisms that might be operating, including those that are unknown, how can specified complexity dispense with them?

Specified complexity can dispense with unknown material mechanisms provided there are independent reasons for thinking that explanations based on known material mechanisms will not be overturned by yet-to-be-identified unknown mechanisms. Such independent reasons typically take the form of arguments from contingency that invoke numerous degrees of freedom. Sometimes they take the form of arguments from exhaustion: after repeatedly trying to achieve a result (say, transforming lead into gold), researchers become convinced that a result can't be achieved—period. Often additional theoretical grounds help strengthen the argument from exhaustion. Alchemy, for instance, was largely discarded before chemistry provided solid theoretical grounds for its rejection. But the rise of modern chemistry, with its theory of elements, put paid to alchemy. (Of course, by alchemy I mean the transformation en masse of a substance by crude interventions like heat applications and potions and not its atom-by-atom reconstruction by sophisticated gadgets like particle accelerators.)

In any event, we need a solid handle on the relevant probability distributions before we can attribute specified complexity with any confidence. Whether we can get such a handle must be assessed on a case-by-case basis—a point I emphasized in *The Design Inference*. There I contrasted the outcome of an agricultural experiment with the opening of a combination lock. With the agricultural experiment, we simply didn't have enough knowledge about the underlying mechanisms that might render one fertilizer better than another for a given crop. Combination locks, however, are a different story. Known material mechanisms (in this case, the laws of physics) prescribe two possible motions of the combination lock: clockwise and counterclockwise rotations. Material mechanisms, however, cannot prescribe the exact turns that open the lock. The geometry and symmetry of the lock preclude that material mechanisms can distinguish one combination from another. That is, one is as good as any other from the vantage of material mechanisms.

Combination locks exhibit numerous degrees of freedom in their possible combinations. In fact, it's precisely these degrees of freedom that guarantee the security of the lock. The more degrees of freedom, the more possible combinations and the more secure the lock. Material mechanisms are compatible with these degrees of freedom and tell us that each possible combination is physically realizable. But precisely because each possible combination is physically realizable, material mechanisms as such cannot mandate one combination to the exclusion of others. For that, we need initial and boundary conditions. These describe the precise circumstances within which material mechanisms may act. For instance, whether dialing a combination lock in a particular way opens it depends on how the tumblers are arranged (i.e., their initial condition).

Thus, to establish that no material mechanism explains a phenomenon, one typically establishes that it is compatible with the known material mechanisms involved in its production but that these mechanisms also permit any number of alternatives to it. (That is, they permit a wide range of initial and boundary conditions, which constitute the degrees of freedom under which the material mechanisms may operate.) By being compatible with but not required by the known material mechanisms involved in its production, a phenomenon becomes irreducible not only to the known mechanisms but also to any unknown mechanisms. How so? Because known material mechanisms can tell us conclusively that a phenomenon is contingent and that it allows full degrees of freedom. Any unknown mechanism would then have to respect that contingency and allow for the degrees of freedom already discovered.

Michael Polanyi described this method for establishing contingency via degrees of freedom in the 1960s. He employed this method to argue for the irreducibility of biology to physics and chemistry. The method applies quite generally: the position of Scrabble pieces on a Scrabble board is irreducible to the natural laws governing the motion of Scrabble pieces, the configuration of ink on a sheet of paper is irreducible to the physics and chemistry of paper and ink, the sequencing of DNA bases is irreducible to the bonding affinities between the bases, and so on.

By establishing a range of possibilities on the basis of known material mechanisms, this method precludes unknown material mechanisms from constricting that range. Scrabble pieces, for instance, can be sequenced in all possible arrangements. For an unknown material mechanism to preclude or prefer some arrangement, it must be suitably constrained by boundary conditions. But then these boundary conditions must in turn allow at least as many degrees of freedom as the possible arrangements of Scrabble pieces (otherwise there wouldn't be complete freedom in the sequencing of Scrabble pieces, which we know there is). It's this regress from the output of material mechanisms to their boundary-condition input that demonstrates the inadequacy of material mechanisms to originate specified complexity. At best, material mechanisms can shuffle around preexisting specified complexity embedded in initial and boundary conditions.

In conclusion, the reliability of specified complexity as a criterion for detecting design must be understood in relation to all material mechanisms that might be operating in a given circumstance. In other words, the criterion is reliable at detecting design, provided all material mechanisms that might be operating in the given circumstance are eliminated. Nevertheless, in practice we can only eliminate the material mechanisms we know about. Although the possible divergence between known mechanisms and all mechanisms *überhaupt* (known and unknown) may seem to undercut the complexity-specification criterion, it really doesn't. If, for instance, there are independent reasons why the probability distributions induced by known mechanisms are secure against unknown mechanisms, then the criterion sweeps the field of all mechanisms that could preclude design.

In thereby eliminating all material mechanisms, we are not saying that a phenomenon is inherently unexplainable. Rather, we are saying that material mechanisms don't explain it and that design does. This conclusion of design derives not from an overactive imagination but simply from following the logic of induction where it leads: In cases where the underlying causal history is known, specified complexity does not occur without design. Specified complexity, therefore, provides inductive support not merely for inexplicability in terms of material mechanisms but also for explicability in terms of design. (And let me stress again, we are not dealing here with an either-or, pitting material mechanisms against design. Rather it is a question of one-or-both, pitting material mechanisms taken in isolation against material mechanisms working in tandem with design.)

13

OBJECTIVITY AND SUBJECTIVITY

Does specified complexity describe an objective feature of the world or merely a subjective state of ignorance about the functioning of the world?

SPECIFIED COMPLEXITY IS A PROPERTY that things can possess or fail to possess. Yet in what sense is specified complexity a property? Properties come in different varieties. There are objective properties that obtain irrespective of who attributes them. Solidity and fluidity are such properties. Water at room temperature is fluid. Water below zero degrees Celsius is solid. Such attributions are perfectly objective. On the other hand, there are also subjective properties that depend crucially on who attributes them. Beauty is such a property. To be sure, beauty may not be entirely in the eye of the beholder (there may be objective aspects to it), but beauty cannot make do without the eye of some beholder.

The distinction between objective and subjective properties has a long tradition in philosophy. With René Descartes, that distinction became important also in science. Descartes made this distinction in terms of primary and secondary qualities. For Descartes, material objects had one primary quality: extension. The other properties of matter—its color or texture, for instance—were secondary qualities that simply described how matter, due to the various ways it was configured or extended, affected our senses. Descartes's distinction between primary and secondary qualities has required updating in light of modern physics. Color, for instance, is nowadays treated as the wavelength of electromagnetic radiation and regarded as a primary quality (though the subjective experience of color is still regarded as a secondary quality). Even so, the idea that some properties are primary or objective and others are secondary or subjective remains with us, especially in the sciences.

The worry, then, is that specified complexity may be entirely a subjective property, with no way of grasping nature at its ontological joints and thus no way of providing science with a valid tool for inquiry. This worry is misplaced but needs to be addressed. The first thing we need to see is that the objective-subjective distinction is not as neat as we might at first think. Consider the following three properties: *X is solid*, *X is married*, and *X is beautiful*. (The *X* here denotes a placeholder to which the properties apply.) The property *X is solid*, as already noted, is objective. Anybody around the world can take a sample of some item, subject it to a chemical test, and determine whether the bonds between the atoms and molecules render it a solid (as opposed to a fluid, gas or plasma). On the other hand, *X is beautiful* seems thoroughly subjective. Even if objective standards of beauty reside in the mind of a cosmic designer or in a Platonic heaven, in practice, people's assessments of beauty differ drastically. Indeed, no single object seems universally admired as beautiful. If specified complexity is subjective in the same way that beauty is, then specified complexity cannot be a useful property for science.

But what about *X is married?* It certainly is an objective fact about the world, whether you or I are married. And yet there is an irreducibly subjective element to this property as well: unlike the solidity of rocks, say, which is simply a fact about nature and does not depend on human subjects, marriage is a social institution that depends intimately on human subjects. Whereas solidity is purely objective and beauty purely subjective, marriage is at once objective and subjective. This confluence of objectivity and subjectivity for social realities like money, marriage and mortgages is the topic of John Searle's *The Construction of Social Reality*. Social realities are objective in the sense that they command intersubjective agreement and express facts (rather than mere opinions) about the social world we inhabit. But they exist within a social matrix, which in turn presupposes subjects and therefore entails subjectivity.

Searle therefore supplements the objective-subjective distinction with an ontological-epistemic distinction (*ontology* referring to what exists, *epistemology* referring to what we know).

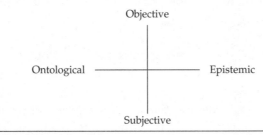

Objective

Ontological ———+——— Epistemic

Subjective

Figure 2 John Searle's social realities

Accordingly, solidity is ontologically objective: it depends on the ontological state of nature and is irrespective of humans or other subjects. Alternatively, beauty is epistemically subjective: it depends on the epistemic state of humans or other subjects, and its assessment is free to vary from subject to subject. Properties reflecting social realities like money, marriage and mortgages, on the other hand, are ontologically subjective but epistemically objective. Thus marriage is ontologically subjective in that it depends on the social conventions of human subjects. At the same time, marriage is epistemically objective: any dispute about somebody being married can be objectively settled on the basis of those social conventions.

How do Searle's categories apply to specified complexity? They apply in two parts, corresponding to the two parts that make up specified complexity. Specified complexity involves a specification, a pattern conditionally independent of some observed outcome. Specified complexity also involves an assignment of complexity (improbability) to the event associated with that pattern. Think of an arrow landing in a target. The target is an independently given pattern and therefore a specification. But the target also highlights an event—the arrow landing in the target—and that event has a certain probability.

Specifications, by being conditionally independent of the outcomes they describe, are, within Searle's scheme, epistemically objective. Moreover, once a specification is given and the event it represents is identified, the probability of that outcome is ontologically objective. Consider, for instance, a quantum mechanical experiment in which polarized light is sent through a polarizing filter whose angle of polarization is at 45 degrees with that of the light. Imagine that the light is sent through the filter photon by photon. According to quantum mechanics, the probability of any photon getting through the filter is 50 percent, and each photon's probability of getting through is probabilistically independent of the others. This quantum mechanical experiment therefore models the flipping of a fair coin (heads = photon passes through the filter; tails = photon doesn't pass through the filter), though without the possibility of any underlying determinism undermining the randomness (assuming quantum mechanics delivers true randomness).

Suppose now that we represent a photon passing through the filter with a one (instead of with heads) and a photon not passing through the filter with a zero (instead of with tails). Consider the specification 1101110111110111111 . . ., namely, the sequence of prime numbers in unary notation. (Successive ones separated by a zero represent each number in the sequence.) For definiteness let's consider the prime numbers between 2 and 101. This representation of prime numbers is ontologically subjective in the sense that it depends on human subjects who know about arithmetic (and specifically about prime numbers and unary notation). It is also epistemically objective inasmuch as arithmetic is a universal aspect of rationality. Moreover, once this specification of primes is in place, the precise probability of a sequence of photons passing through the filter and matching it is ontologically objective. Indeed, that probability will depend solely on the inherent physical properties of photons and polarizing filters. Specified complexity, therefore, is at once epistemically objective (on the specification side) and ontologically objective (on the complexity side once a specification is in hand).

Specified complexity therefore avoids the charge of epistemic subjectivity, which, if true, would relegate specified complexity to the whim, taste or opinion of subjects. Yet specified complexity does not merely avoid this charge. More positively, it also displays two desirable forms of objectivity: specifications are epistemically objective, and measures of complexity based on those specifications are ontologically objective. Is this enough to justify specified complexity as a legitimate tool for science? To answer that question, let's consider what could go awry with specified complexity to prevent it from functioning as a legitimate tool within science.

According to David Berlinski, specifications are the problem. In the December 2002 issue of *Commentary* in an article titled "Has Darwin met his match?" he wrote,

> A specification is a human gesture. It may be offered or withheld, delayed or deferred; it may be precise or incomplete, partial or unique. Paley's watch may thus be specified in terms of its timekeeping properties, but it may also be specified in terms of the arrangement or number of its parts; and its parts may be specified in horological, mechanical, molecular, atomic, or even sub-atomic terms. Each specification introduces a different calculation of probabilities. Paley's watch may be improbable under one specification, probable under another.

A gesture is an expression of thought that focuses attention on one thing to the exclusion of another. Fine, let's grant that specifications are gestures. It doesn't follow that all gestures are created equal or that none may be legitimately employed in drawing design inferences. Berlinski's criticism spuriously homogenizes all gestures. If all I do with Paley's watch (see

chapter seven) is gesture at the number of its parts, then no design inference will be forthcoming. On the other hand, if I gesture at its functional
complexity and integration of parts for the purpose of telling time, then a
design inference will be forthcoming.

Berlinski is concerned that different specifications may yield different
probability assessments. Thus, one specification might warrant a design
inference but another may not. Yet why should this fact about specifications undercut the ability of specified complexity to elicit a valid design inference? Imagine two superimposed targets with a single arrow lodged in
both bull's-eyes. Suppose the bull's-eye of one target is huge and that of
the other is tiny (implying that the arrow landing in the large bull's-eye is
reasonably probable but for the small bull's-eye is minuscule). Suppose
further that the targets were positioned independently of the arrow's trajectory. In that case we would infer design as responsible for the arrow's
trajectory. Indeed, the arrow lodged in the small bull's-eye would exhibit
specified complexity and therefore trigger a design inference. As for the arrow lodged in the large bull's-eye, its failure to exhibit specified complexity is irrelevant. To draw a successful design inference, it is enough to find
at least one specification for which the item in question exhibits specified
complexity. (The failure of other specifications is beside the point.) Nor
does this mean that regardless of the item in question, we can always find
a specification for which it exhibits specified complexity. Specified complexity requires specifications of low specificational complexity, and these
are few and far between (recall chapter ten).

Specifications are not the problem. True, specifications, though epistemically objective, are not ontologically objective. The failure of specifications
to be ontologically objective, however, does not prevent them from playing a legitimate role in the natural sciences. In biology, specifications are
independently given functional patterns that describe the goal-directed
behavior of biological systems. A bacterial flagellum, for instance, is a bidirectional motor-driven propeller on the backs of certain bacteria that
moves them through their watery environments. This functional description is epistemically objective, but on any naturalistic construal of science
it must be regarded as ontologically subjective. If nature, as naturalism requires, is a closed nexus of undirected natural causes, then nature knows
nothing about such functional descriptions. And yet biology as a science
would be impossible without them. Indeed, the very concepts of survival
and reproduction, on which evolutionary theory hinges, are themselves
such functional descriptions. Functional language is indispensable to biol-

ogy. Specifications clarify and make precise that language.

Any problem justifying specified complexity's legitimacy within science therefore resides elsewhere. Indeed, if there is a problem, it resides with complexity. Although complexity becomes ontologically objective once a specification is in place, our assessment of complexity is just that—our assessment. And the problem with assessments is that they can be wrong. Specifications are under our control. We formulate specifications on the basis of background knowledge. The complexity denoted by specified complexity, on the other hand, resides in nature. This form of complexity corresponds to a measure of probability, and such probabilities depend on the way nature is constituted. There is an objective underlying fact as to what these probabilities are. But our grasp of these probabilities can be less than adequate. The problem, then, with finding specified complexity an effective place in science is to create a bridge between complexity as it exists in nature and our assessments of that complexity. Alternatively, the problem is not whether specified complexity is a coherent scientific property but whether we can justify particular attributions of specified complexity to particular objects or events in nature. Specified complexity may be fine as a theoretical construct, but can we actually apply it in practice? That question brings us to our next topic—assertibility.

14

ASSERTIBILITY

*Even if specified complexity is a
well-defined, objective and reliable
criterion for detecting design, why should
we think that we could ever be justified
in asserting that some natural object
exhibits specified complexity?*

THIS QUESTION LEAVES SPECIFIED COMPLEXITY'S coherence, objectivity
and reliability as a criterion for detecting design unchallenged (and appropriately so, since, as we've seen in the preceding chapters, these are all unproblematic). Rather, the concern here is that determining whether something exhibits specified complexity is impracticable, like reconstructing
something from the past for which all the relevant evidence has been lost.
Did Julius Caesar have a mole on the sole of his left foot? The question is
meaningful, and there is an objective underlying fact that correctly settles
this question. (Caesar either did or didn't have that mole.) Nevertheless,
all the evidence that might help us resolve this question is long lost. So too
with specified complexity, the worry is that it may be a meaningful concept in the abstract but that there's no way to apply it practically to settle
questions about design. In other words, the worry here centers on specified complexity's *assertibility.*

Assertibility (spelled with an *i*) is a philosophical term that refers to the
epistemic or rational justification for a claim. It is distinguished from *assertability* (spelled with an *a*), which refers to the local factors that in the
pragmatics of discourse determine whether asserting a claim is appropriate. For instance, as a tourist in Iraq prior to the 2003 war with the United
States, I would have been epistemically justified asserting that Saddam
Hussein was a monster. (I would have had plenty of compelling evidence
to justify that claim, making it assertible.) Local-pragmatic considerations,
however, would have told against asserting such a remark inside Iraq. (By

uttering the claim there, I would have violated societal and political expectations; consequently, the claim there would have been unassertable.) Unlike assertibility, assertability can depend on anything from etiquette and good manners to who happens to hold political power. Assertibility (with an *i*) is what interests us here—whether a claim can be rationally justified, regardless of how it plays with our other sensibilities or expectations.

To illustrate what's at stake in regard to specified complexity, consider an analogy from mathematics. There exist numbers whose decimal expansions are such that each digit between zero and nine has relative frequency exactly 10 percent as the decimal expansion becomes arbitrarily large (or, as mathematicians would say, the relative frequency of each digit *in the limit* is exactly 10 percent). The simplest such number is perhaps .01234567890123456789 . . . where the pattern 0123456789 just keeps repeating. Let's call such numbers *regular*. (Mathematicians typically prefer a stronger notion of regularity called *normality*, which characterizes the limiting behavior of all finite strings of digits and not merely that of single digits. For the purposes of this example, however, the concept *regularity* suffices.) Thus, the property *X is regular* applies to this number. Regularity is clearly a legitimate mathematical property: it is perfectly well defined, and numbers either are regular or fail to be regular.

But suppose next we want to determine whether the number pi is regular. (Pi equals the ratio of the circumference of a circle to its diameter.) Pi has a nonrepeating decimal expansion. Over the years mathematicians and computer scientists have teamed up to compute as many digits of pi as mathematical methods and computer technology permit. The current record stands at 206,158,430,000 decimal digits of pi and is held by the Japanese researchers Yasumasa Kanada and Daisuke Takahashi. (The currently standard 40 gigabyte hard drive is too small to store this many decimal digits.) Among these 200 billion decimal digits of pi, each of the digits between zero and nine has relative frequency roughly 10 percent. Is pi therefore regular?

Just as there is an underlying physical fact whether an object or event in nature exhibits specified complexity, so there is an underlying mathematical fact whether pi is regular. Pi either is regular or fails to be regular. Nonetheless, the determination whether pi is regular is another matter. With the number .01234567890123456789 . . ., its regularity is evident by inspection. But the decimal digits of pi are nonrepeating, and to date there is no theoretical justification of its regularity. The closest thing to a justification is to point out that for the standard probability measure on the unit interval (known as Lebesgue measure), all numbers except for a set of probability zero are regular. The presumption, then, is that pi is likely to be regular. The problem with such a "probabilistic proof" of pi's regularity, however, is that we have no basis, mathematical or otherwise, for thinking that pi was sampled according to Lebesgue probability measure and therefore is likely to fall among the regular numbers. In short, mathematical theory does not justify asserting that pi is regular.

Nor does mathematical experience help. Even the discovery that the single digits of pi have approximately the right relative frequencies for pi's first 200 billion decimal digits provides no basis for confidence that pi is regular. However regular the decimal expansion of pi looks in some initial segment, it could go haywire thereafter, possibly even excluding certain single digits entirely after a certain point. On the other hand, however nonregular the decimal expansion of pi looks in some initial segment, the relative frequencies of the single digits between zero and nine could eventually settle down into the required 10 percent, and pi itself could be regular (any initial segment thereby getting swamped by the infinite decimal expansion that lies beyond it). Thus, to be confident that pi is regular, mathematicians need a strict mathematical proof showing that each single digit between zero and nine has a limiting relative frequency of exactly 10 percent.

Now critics of intelligent design demand this same high level of justification (i.e., mathematical proof) before they accept specified complexity as a legitimate tool for science. Yet a requirement for strict proof, though legitimate in mathematics, is entirely wrong-headed in the natural sciences. The natural sciences make empirically based claims, and such claims are always falsifiable. (Even Newtonian mechanics, which for a time defined physics, ended up being falsified.) Errors in measurement, incomplete knowledge, limited theoretical insight and the problem of induction cast a shadow over all scientific claims. To be sure, the shadow of falsifiability doesn't incapacitate science. But it does make the claims of science (unlike those of mathematics) tentative, and it also means that we need to pay special attention to how scientific claims are justified.

A little reflection makes clear that this attempt by skeptics to undo specified complexity cannot be justified on the basis of scientific practice. Indeed, the skeptic imposes requirements so stringent that they are absent from every other aspect of science. If standards of scientific justification are set too high, no interesting scientific work will ever get done. Science therefore balances its standards of justification with the requirement for self-correction in light of further evidence. The possibility of self-correction in light of further evidence is absent in mathematics and accounts for mathematics' need for the highest level of justification, namely, strict logico-deductive proof. But science does not work that way.

The key question for this discussion, therefore, is how to justify ascribing specified complexity to natural structures. To see what's at stake, consider further the analogy between the regularity of numbers and the specified complexity of natural structures. We need to be clear where that analogy holds and where it breaks down. The analogy holds insofar as both specified complexity and regularity make definite claims about some objective underlying fact. In the case of regularity, it is an underlying mathematical fact: the decimal expansions of numbers either exemplify or fail to exemplify regularity. In the case of specified complexity, it is an underlying physical fact: a biological system, for instance, either exemplifies or fails to exemplify specified complexity. This last point is worth stressing. Attributing specified complexity is never a meaningless assertion. On the assumption that no design or teleology was involved in the production of some event, that event has a certain probability based on natural causal mechanisms. That probability in turn maps onto an associated measure of complexity (see chapter ten). Whether the level of complexity is high enough to qualify the event as exemplifying specified complexity depends on the physical conditions surrounding the event. In any case, there is a definite fact of the matter whether specified complexity obtains.

Any problem with ascribing specified complexity to that event therefore resides not in its coherence as a meaningful concept: specified complexity is well-defined. If there is a problem, it resides, as we noted earlier, in what philosophers call its *assertibility*. Assertibility, as defined above, refers to our justification for asserting the claims we make. A claim is assertible if we are justified asserting it. With the regularity of pi, it is possible that pi is regular. Thus in asserting that pi is regular, we might be making a true statement. But without a mathematical proof of pi's regularity, we have no justification for asserting that pi is regular. The regularity of pi is, at least for now, unassertible. But what about the specified complexity of various biological systems? Are there any biological systems whose specified complexity is assertible?

Critics of intelligent design argue that no attribution of specified complexity to any natural system can ever be assertible. The argument starts by noting that if some natural system exemplifies specified complexity, then it must be vastly improbable with respect to all purely natural mechanisms that could be operating to produce it. But that means calculating a probability for each such mechanism. This, so the argument runs, is an impossible task. At best science could show that a given natural system is vastly improbable with respect to known mechanisms operating in known ways and for which the probability can be estimated. But that omits, first, known mechanisms operating in known ways for which the probability cannot be estimated; second, known mechanisms operating in unknown ways; and, third, unknown mechanisms.

Thus, even if it is true that some natural system exemplifies specified complexity, we could never legitimately assert its specified complexity, much less know it. Accordingly, to assert the specified complexity of any natural system constitutes an argument from ignorance (see chapters thirty and thirty-one). This line of reasoning against specified complexity is much like the standard agnostic line against theism: we can't prove atheism (cf. the total absence of specified complexity from nature), but we can show that theism (cf. the specified complexity of certain natural systems) cannot be justified and is therefore unassertible. This is how skeptics argue that there is no—and indeed can be no—evidence for God or design.

The problem with this line of reasoning is that science must work with available evidence and on that basis (and that basis alone) formulate the best explanation of the phenomenon in question. This means that science cannot explain a phenomenon by appealing to the promise, prospect or possibility of future evidence. In particular, unknown mechanisms or undiscovered ways by which those mechanisms operate cannot be invoked to explain a phenomenon. If known material mechanisms can be shown incapable of explaining a phenomenon, then it is an open question whether any mechanisms whatsoever are capable of explaining it. If, further, good reasons exist for asserting the specified complexity of certain biological systems, then design itself becomes assertible in biology.

Take, for instance, the bacterial flagellum. Despite the thousands of research articles on it, no mechanistic account of its origin exists. Consequently, there is no evidence against its being complex and specified. It is therefore a live possibility that it is complex and specified. But is it fair to *assert* that it is complex and specified—in other words, to *assert* that it exhibits specified complexity? The bacterial flagellum is irreducibly complex, meaning that all its components are indispensable for its function as a motility structure. What's more, it is minimally complex, meaning that any structure performing the bacterial flagellum's function as a bidirectional motor-driven propeller cannot make do without its basic components.

Consequently, no direct Darwinian pathway exists that incrementally adds these basic components and therewith evolves a bacterial flagellum. Rather, an indirect Darwinian pathway is required, in which precursor systems performing different functions evolve by changing functions and components over time. (Darwinists refer to this as *coevolution* and *co-optation*.) Plausible as this sounds (to the Darwinist), there is no convincing evidence for it (see chapters thirty-eight and forty). What's more, evidence from engineering strongly suggests that tightly integrated systems like the bacterial flagellum are not formed by trial-and-error tinkering in which form and function coevolve. Rather, such systems are formed by a unifying conception that combines disparate components into a functional whole—in other words, by design.

Does the bacterial flagellum exhibit specified complexity? Is such a claim assertible? Certainly the bacterial flagellum is specified. One way to see this is to note that humans developed bidirectional motor-driven propellers well before they figured out that the flagellum was such a machine. This is not to say that for the biological function of a system to constitute a specification, humans must have independently invented a system that performs the same function. Nevertheless, independent invention makes all the more clear that the system satisfies independent functional requirements and therefore is specified. At any rate, no biologist I know questions whether the functional systems that arise in biology are specified. At issue always is whether the Darwinian mechanism, by employing natural selection, can overcome the vast improbabilities that at first blush seem to arise with such systems. To overcome a vast improbability, the Darwinian mechanism attempts to break it into a sequence of more manageable probabilities.

To illustrate what's at stake in this dividing and conquering of improbabilities, suppose a hundred pennies are tossed. What is the probability of getting all one hundred pennies to exhibit heads? The probability depends on the chance process by which the pennies are tossed. If, for instance, the chance process operates by tossing all the pennies simultaneously and does not stop until all the pennies simultaneously exhibit heads, it will require on average about a thousand billion billion billion such simultaneous tosses for all the pennies to exhibit heads in one of those tosses. If, on the other hand, the chance process tosses only those pennies that have not yet exhibited heads, then after about eight tosses, on average, all the pennies will exhibit heads. Darwinists tacitly assume that all instances of biological complexity are like the second case, in which a seemingly vast improbability can be broken into a sequence of reasonably probable events by gradually improving on an existing function. (In the case of our pennies, improved function corresponds to exhibiting more heads.)

Irreducible and minimal complexity challenge the Darwinian assumption that vast improbabilities can always be broken into manageable probabilities. What evidence there is suggests that such instances of biological complexity must be attained simultaneously (as when the pennies are tossed simultaneously). In such cases, gradual Darwinian improvement offers no help in overcoming their improbability. Thus, when we analyze structures like the bacterial flagellum probabilistically on the basis of known material mechanisms operating in known ways, we find that they are highly improbable and therefore complex in the sense required by specified complexity.

Is it therefore legitimate to assert that the bacterial flagellum exhibits specified complexity? Design theorists say yes. Evolutionary biologists say no. As far as they are concerned, design theorists have failed to take into account indirect Darwinian pathways by which the bacterial flagellum might have evolved through a series of intermediate systems that changed function and structure over time in ways that we do not yet understand. But is it that we do not yet understand the indirect Darwinian evolution of the bacterial flagellum or that it never happened that way in the first place? At this point there is simply no convincing evidence for such indirect Darwinian evolutionary pathways to account for biological systems that display irreducible and minimal complexity. All that Darwinists have done until now is identify subsystems of the bacterial flagellum that could serve some biological function of their own—much as the motor of a motorcycle might serve some function on its own (perhaps as a heater or blender). But there's nothing exceptional here for design; designed systems that perform one function are typically made up of designed subsystems that have their own function.

Is this, then, where the debate ends—with evolutionary biologists chiding design theorists for not working hard enough to discover those (unknown) indirect Darwinian pathways that lead to the emergence of irreducibly and minimally complex biological structures like the bacterial flagellum? Alternatively, does it end with design theorists chiding evolutionary biologists for deluding themselves that such indirect Darwinian pathways exist when all the available evidence suggests that they do not?

Although this may seem like an impasse, it really isn't. Like compulsive gamblers who are constantly hoping that some big score will cancel their debts, evolutionary biologists live on promissory notes that show no sign of being redeemable. As noted before, science must form its conclusions on the basis of available evidence, not on the possibility of future evidence. If evolutionary biologists can discover or construct detailed, testable, indirect Darwinian pathways that account for the emergence of irreducibly and minimally complex biological systems like the bacterial flagellum, then more power to them—intelligent design will quickly pass into oblivion. But until that happens, evolutionary biologists who claim that natural selection accounts for the emergence of the bacterial flagellum are worthy of no more credence than compulsive gamblers who are forever promising to settle their accounts.

Evolutionary biologists cannot even justify looking to future evidence by pointing to current progress because they have not made any meaningful progress accounting for biological complexity. (If they had, we wouldn't be having this discussion.) Again, they're like compulsive gamblers, owing millions of dollars to the local loan shark, begging for an extension but without being able to point to a short track record of rehabilitation, steady work or even a few hundred dollars of repayment money. What is the origin of biological complexity, and how is it to be explained? For all the insights evolutionary biology has brought to bear on this question, we might just as well return to the state of biology prior to Darwin. Evolutionary biology has no idea whatsoever how to answer this question.

There is further reason to be skeptical of evolutionary biology and side here with intelligent design. In the case of the bacterial flagellum, what keeps evolutionary biology afloat is the possibility of indirect Darwinian pathways that might account for it. Practically speaking, this means that even though no slight modification of a bacterial flagellum can continue to serve as a motility structure, a slight modification could serve some other function. But there is now mounting evidence of biological systems for which any slight modification does not merely destroy the system's existing function but also destroys the possibility of any function of the system whatsoever. (Consult, for instance, the research on extreme functional sensitivity of various enzymes and on irreducibly complex metabolic pathways of enzymes for which each enzyme needs to attain a certain catalytic threshold before it or its associated pathway can serve any biological function at all.) For such systems, neither direct nor indirect Darwinian pathways could account for them. In that case we would be dealing with an in-principle argument showing not simply that no known material mechanism is capable of accounting for the system, but also that any unknown material mechanism is incapable of accounting for it as well. The argument here turns on an argument from contingency and degrees of freedom.

It is possible to rule out unknown material mechanisms once and for all, provided one has independent reasons for thinking that explanations based on known material mechanisms cannot be overturned by yet-to-be-identified unknown mechanisms. Such independent reasons typically take the form of arguments from contingency that invoke numerous degrees of freedom. Thus, to establish that no material mechanism explains a phenomenon, we must establish that it is compatible with the known material mechanisms involved in its production, but that these mechanisms also permit any number of alternatives to it. By being compatible with, but not required by, the known material mechanisms involved in its production, a phenomenon becomes irreducible not only to the known mechanisms but also to any unknown mechanisms. How so? Because known material mechanisms can tell us conclusively that a phenomenon is contingent and allows full degrees of freedom. Any unknown mechanism would therefore have to respect that contingency and allow for the degrees of freedom already discovered.

Consider, for instance, a configuration space comprising all possible character sequences from a fixed alphabet. (Such spaces model not only written texts but also polymers like DNA, RNA and proteins.) Configuration spaces like this are perfectly homogeneous, with one character string geometrically interchangeable with the next. The geometry therefore precludes any underlying mechanisms from distinguishing or preferring some character strings over others. Not material mechanisms but external semantic information (in the case of written texts) or functional information (in the case of biopolymers) is needed to generate specified complexity in these instances. To argue that this semantic or functional information reduces to material mechanisms is like arguing that Scrabble pieces have inherent in them preferential ways they like to be sequenced. They don't. Michael Polanyi made such arguments for biological design in the 1960s. Stephen Meyer has updated them for the present.

So is the claim that the bacterial flagellum exhibits specified complexity assertible? You bet. Science works with available evidence, not with vague promises of future evidence. Our best evidence points to the specified complexity (and therefore design) of the bacterial flagellum. It is therefore incumbent on the scientific community to admit, at least provisionally, that the bacterial flagellum is designed. Nor should opponents of intelligent design comfort themselves with any misplaced notion that the intelligent design movement is and will be powered solely by the bacterial flagellum. Assertibility comes in degrees, corresponding to the strength of evidence that justifies a claim. That the bacterial flagellum exhibits specified complexity is highly assertible—this despite the logical impossibility of ruling out the infinity of possible indirect Darwinian pathways that might give rise to it. Yet for other systems, like enzymes that exhibit extreme functional sensitivity, there could be compelling grounds for ruling out such indirect Darwinian pathways as well. The assertibility for the specified complexity of such systems could therefore prove stronger still.

The evidence for intelligent design in biology is thus destined to grow ever stronger. There's only one way evolutionary biology can defeat intelligent design, and that is by in fact solving the problem that it claimed all along to have solved but in fact never did—to account for the emergence of multipart, tightly integrated complex biological systems (many of which display irreducible and minimal complexity) apart from teleology or design. To claim that the Darwinian mechanism solves this problem is false. The Darwinian mechanism is not itself a solution but rather a template for the type of solution that Darwinists hope can solve the problem. Templates, however, require details, and filling in the details of their template is the one thing Darwinists never do. That's why molecular biologist James Shapiro, who is not a design theorist, writes, "There are no detailed Darwinian accounts for the evolution of any fundamental biochemical or cellular system, only a variety of wishful speculations" (quoted from his 1996 book review of *Darwin's Black Box* that appeared in the *National Review*).

In summary, specified complexity is a well-defined property that can be meaningfully affirmed or denied of events and objects in nature. Specified complexity is an objective property: specifications are epistemically objective, and complexity is ontologically objective (see chapter thirteen). Any concern over specified complexity's legitimacy within science rests not with its coherence or objectivity but with its assertibility—with whether, and the degree to which, ascribing specified complexity to some natural object or event is justified. Any blanket attempt to render specified complexity unassertible gives naturalism an unreasonable advantage, ensuring that design cannot be discovered even if it is present in nature. Whereas naturalism looks to future evidence to overturn intelligent design, science can proceed only on the basis of available evidence. As a consequence, ascriptions of specified complexity to natural objects and events, and to biological systems in particular, can be assertible. And indeed, there are actual biological systems for which ascribing specified complexity—and therefore design—is eminently assertible.

15

THE CHANCE OF THE GAPS

*Why must any scientific theory that aims
to detect design be probabilistic?*

SCIENTISTS RIGHTLY WORRY ABOUT THE GOD OF THE GAPS, in which God is used as a stopgap for ignorance. But chance can play exactly this role also. Science therefore has to be able to eliminate chance when the probability of events gets too small. If not, chance can be invoked to explain anything. For instance, in the movie *This Is Spinal Tap*, the lead singer of the fictional band Spinal Tap remarks that a former drummer died by "spontaneously combusting." Any one of us could this instant spontaneously combust if all the most rapidly moving air molecules in our vicinity suddenly converged on us. Such an event, however, is highly improbable, and no one takes it seriously.

High improbability by itself, however, is not enough to preclude chance. Indeed, highly improbable events happen all the time. Flip a coin a thousand times, and you'll participate in a highly improbable event. Just about anything that happens is highly improbable once we factor in all the ways what did happen could have happened. Mere improbability therefore fails to rule out chance. In addition, improbability needs to be joined with an independently given pattern. An arrow shot randomly at a large blank wall will be highly unlikely to land at any one place on the wall. Yet land somewhere it must, and so some highly improbable event will be realized. But now fix a target on that wall and shoot the arrow. If the arrow lands in the target and the target is sufficiently small, then chance is no longer a reasonable explanation of the arrow's trajectory.

Highly improbable, independently patterned events exhibit what I call *specified complexity.* I have shown that specified complexity is a reliable empirical marker of intelligent agency (see chapter twelve). Nevertheless, a persistent worry about small-probability arguments remains: given an independently given pattern, or specification, what level of improbability must be attained before chance can legitimately be precluded? A wall so large that it cannot be missed and a target so large that covers half the wall, for instance, are hardly sufficient to preclude chance (or "beginner's luck") as the reason for an archer's success hitting the target. The target needs to be small to preclude hitting it by chance.

But how small is small enough? To answer this question we need the concept of a probabilistic resource. A probability is never small in isolation but only in relation to a set of probabilistic resources that describe the number of relevant ways an event might occur or be specified. There are thus two types of probabilistic resources: *replicational* and *specificational*. To see what is at stake, consider a wall so large that an archer cannot help but hit it. Next, let us say we learn that the archer hit some target fixed to the wall. We want to know whether the archer could reasonably have been expected to hit the target by chance. To determine this we need to know any other targets at which the archer might have been aiming. Also, we need to know how many arrows were in the archer's quiver and might have been shot at the wall. The targets on the wall constitute the archer's specificational resources. The arrows in the quiver constitute the archer's replicational resources.

Probabilistic resources comprise the relevant number of ways an event can occur (replicational resources) and be specified (specificational resources). The important question therefore is not, What is the probability of the event in question? but rather, What does its probability become after all the relevant probabilistic resources have been factored in? Probabilities can never be considered in isolation but must always be referred to a relevant reference class of possible replications and specifications. A seemingly improbable event can become quite probable when placed within the appropriate reference class of probabilistic resources. On the other hand, it may remain improbable even after all the relevant probabilistic resources have been factored in. If it remains improbable and if the event is also specified, then it exhibits specified complexity.

In the observable universe, probabilistic resources come in limited supplies. In fact, it can be shown that any specified event of probability less than 1 in 10^{150} will remain improbable even after all conceivable probabilistic resources from the observable universe have been factored in (see chapter ten). A probability of 1 in 10^{150} is therefore a *universal probability bound*. A specified event of probability less than this universal probability bound cannot be rendered reasonably probable even if all available probabilistic resources in the known universe are brought to bear against it.

Implicit in a universal probability bound such as 1 in 10^{150} is that the universe is too small a place to generate specified complexity by sheer exhaustion of possibilities. Stuart Kauffman develops this theme at length in his book *Investigations* (Oxford University Press, 2000). In one of his examples (and there are many like it throughout the book), he considers the number of possible proteins of length 200 (i.e., 20^{200} or approximately 10^{260}) and the maximum number of pairwise collisions of particles throughout the history of the universe (he estimates 10^{193} total collisions supposing the reaction rate for collisions is measured in femtoseconds). Kauffman concludes that the known universe hasn't had time since the big bang to run through all possible proteins of length 200 even once. To emphasize this point, he notes it would take more than 10^{67} times the current time span of the universe to construct all possible proteins of length 200 even once. Kauffman even has a name for numbers that are so big that they are beyond the reach of operations performable by and within the universe—*transfinite*.

Kauffman often writes about the universe being unable to exhaust some set of possibilities. Yet at other times he puts an adjective in front of the word *universe*, claiming it is the *known universe* that is unable to exhaust some set of possibilities. Is there a difference between the *universe* (no adjective in front) and the *known* or *observable universe* (adjective in front)? To be sure, there is no empirical difference. Our best scientific observations tell us that the world surrounding us appears quite limited. Indeed, the size, duration and composition of the known universe are such that 10^{150} is a transfinite number in Kauffman's sense. For instance, if the universe were a giant computer, it could perform no more than this number of operations. (Quantum computation, by exploiting superposition of quantum states, enriches the operations performable by an ordinary computer but cannot change their number.) If the universe were devoted entirely to generating specifications, this number would set an upper bound. If cryptographers confine themselves to brute-force methods to test cryptographic keys, the number of keys they can test will always be less than this number.

But what if the universe is in fact much bigger than the known universe? What if the known universe is but an infinitesimal speck within the actual universe? Alternatively, what if the known universe is but one of many universes, each of which is as real as the known universe but causally inaccessible to it? If so, are not the probabilistic resources needed to eliminate chance vastly increased, and is not the validity of 1 in 10^{150} as a universal probability bound thrown into question? This line of reasoning has gained widespread currency among scientists and philosophers in recent years. But in fact, this line of reasoning is fatally flawed. It is illegitimate to rescue chance by invoking probabilistic resources from outside the known universe. To do so artificially inflates one's probabilistic resources.

Only probabilistic resources from the known universe may legitimately be employed in evaluating chance hypotheses. In particular, probabilistic resources imported from outside the known universe are incapable of overturning the universal probability bound of 1 in 10^{150}. The rationale for this claim is straightforward: it is never enough to postulate probabilistic resources merely to prop an otherwise failing chance hypothesis; rather, one needs independent evidence whether there really are enough probabilistic resources to render chance plausible.

Consider, for instance, a state lottery. Suppose we know nothing about the number of lottery tickets sold and are informed simply that the lottery had a winner. Suppose further that the probability of any lottery ticket producing a winner is extremely low. What can we conclude? Does it follow that many lottery tickets were sold? Not at all. We are entitled to this conclusion only if we have independent evidence that many lottery tickets were sold. Apart from such evidence we have no way of assessing how many tickets were sold, much less whether the lottery was conducted fairly and whether its outcome was due to chance. It is illegitimate to take an event, decide for whatever reason that it must be due to chance, and then propose numerous probabilistic resources because otherwise chance would be implausible. I call this the *inflationary fallacy*.

The inflationary fallacy underlies a number of proposals by physicists and philosophers to vastly increase the size of the known universe. These include the bubble universes of Alan Guth's inflationary cosmology, the many worlds of Hugh Everett's interpretation of quantum mechanics, the self-reproducing black holes of Lee Smolin's cosmological natural selection and the possible worlds of David Lewis's extreme modal realist metaphysics. Each of these proposals purports to resolve some problem of general interest and importance in science or philosophy. The details of these proposals are not important here. What is important is that none of them possesses independent evidence for the existence of the entity or process proposed.

Independent evidence helps establish a claim apart from any appeal to its explanatory virtue. The demand for independent evidence is a necessary constraint on theory construction in science so that theory construction does not degenerate into total free play of the mind. My favorite story illustrating the interplay between independent evidence and explanatory virtue is due to John Leslie (*Universes*, 1989). Suppose an arrow is fired at random into a forest and hits Mr. Brown. To explain such a chance occurrence it would suffice for the forest to be full of people. The hypothesis that the forest is full of people therefore possesses explanatory virtue. Even so, this explanation remains but a speculative possibility until it is supported by independent evidence of people other than Mr. Brown actually being in the forest.

Yet with the proposals to inflate the known universe, no such independent evidence is forthcoming. Worse yet, no such independent evidence can be forthcoming. Each of these proposals entails a universe that is effectively infinite (though the portion accessible to us is quite finite). Now the problem with an infinite universe is that human investigators can have no empirical access to its infinity. Indeed, our sensory system is capable of delivering only so many experiences. Consider that a digital video disk (DVD) contains no more than about 10 gigabytes of data (i.e., 10^{10} bytes of information) but can reasonably capture two hours of a human being's visual and auditory experience. It's probably safe to say that a human being's entire sensory experience for one hour (taste, touch and smell in addition to sight and sound) can be captured with a high degree of accuracy and resolution by one petabyte (i.e., 10^{15} bytes). We can think of one petabyte as the equivalent of running a 100,000 DVDs simultaneously. Surely that's enough to cover the range of our sensory experiences for a single hour.

Now the average human life span is less than one hundred years. With twenty-four hours in a day and 365 days in a year, that means humans have less than a million hours in which to live their lives. (There are 876,000 hours in a 100-year life not counting leap years, which add another 600 hours.) It follows that the entire sensory experience of a human being can be captured in one zettabyte (i.e., 10^{21} bytes). Any scientific theory that is the product of a single human scientist will therefore have to be made on the basis of no more than one zettabyte of information. Any scientific theory that is the product of a community of N human scientists will therefore have to be made on the basis of N zettabytes of information. Now the only obligation of an empirically adequate scientific theory is that it be faithful to these few zettabytes of information. Thus, a scientific theory that posits an infinite universe necessarily exceeds anything that's empirically warranted. Call it physics untethered to observation or call it metaphysics, it doesn't matter. The infinite is beyond empirical observation, which means that any appeal to the infinite in our scientific theories signifies not that our finite experience has given us a window on the infinite but rather that we are using infinity as a construct to approximate our finite experience. (As Peter Huber at MIT used to say, "We use the infinite to approximate the finite.")

The only way around these strong finiteness limitations on human experience is for humans to transcend their biology. Christian theology holds such a promise by resurrecting and thereby transforming our physical bodies into spiritual bodies (see 1 Corinthians 15). The materialist, however, doesn't have that option. Confined to understanding all of reality in terms of material mechanisms, the best the materialist can do is merge humans with machines and thereby increase human sensory and processing capacities. The most radical of these proposals is that we upload ourselves onto a superdupercomputer, preferably a quantum computer, and thus dispense with our biology entirely (see Ray Kurzweil's *The Age of Spiritual Machines*).

But there are two problems with this proposal. First, there's no evidence that consciousness and the sensory experience that goes with it has anything to do with complexity or computation. To be sure, on the assumption of materialism, consciousness must reduce to complexity. But consciousness remains a mystery for materialism. Biological survival and reproduction could make do quite nicely without it. It's easy to imagine a world of robot creatures doing all the right things without consciousness. The other problem is that machines, even the fastest and biggest superduper quantum parallel processors, are still finite. I argued that a human being's sensory experiences throughout an entire lifetime could be captured in one zettabyte of information. But even if the entire known universe were a computer, it could never perform more than 10^{150} elementary calculations (for the same reason that the universal probability bound is 1 in 10^{150}). What one means by an elementary computation differs between a conventional and a quantum computer, but there is no escaping the finiteness of computation, whatever form it takes. At no point in such a computer's existence will anything but a finite number of items of information be stored in memory and a finite number of processing steps be executed.

Thus we see that an infinite universe cannot even in principle admit independent evidence. But perhaps an infinite universe's explanatory virtue offsets its inability to admit independent evidence. So what if an infinite universe cannot be grounded in an empirically based physics. It can certainly be posited as a metaphysical hypothesis. Indeed, as a metaphysical hypothesis it increasingly is doing a lot of work, not least defeating any form of transcendent design. An infinite universe underwrites unlimited probabilistic resources, and these in turn allow us to dispense with design in nature. Indeed, unlimited probabilistic resources allow us to explain absolutely everything by reference to chance—not just natural objects that actually did result by chance and not just natural objects that look designed, but also artificial objects that are in fact designed. And here I don't mean we explain away an artificial object by saying that the designer was merely a coincidence of atoms and energy, environment, and genetics, which caused the designer's brain and body parts to move in such a way as to bring about the artificial object. That's precisely the point at issue. In any case, we are still dealing with a designer who, from at least a practical standpoint, consciously set out to create the artificial object.

The idea of an infinite number of causally separate universes allows for an even more bizarre possibility. Was Artur Rubinstein a great pianist, or was it just that whenever he sat at the piano, he happened by chance to put his fingers on the right keys to produce beautiful music? It could happen by chance, and there is some corner of an infinite universe (i.e., some possible world) where everything is exactly as it is in our world except that the counterpart to Rubinstein cannot read or even appreciate music and happens to be incredibly lucky whenever he sits at the piano (like a monkey who, on a single try, randomly types out Milton's *Paradise Lost*). Examples like this can be multiplied. There are possible worlds in which I cannot do arithmetic and yet sit down at my computer and write probabilistic tracts about intelligent design. Perhaps Shakespeare was a genius. Perhaps Shakespeare was an imbecile who just by chance happened to string together a long sequence of apt phrases. Unlimited probabilistic resources ensure not only that we will never know but also that we have no rational basis for preferring one possibility to the other.

Not so fast. Given unlimited probabilistic resources, there does appear to be one way to rebut such anti-inductive skepticism, and that is to admit that while unlimited probabilistic resources allow bizarre possibilities like this, these possibilities are nonetheless highly improbable in the little patch of reality that we inhabit. Unlimited probabilistic resources make bizarre possibilities unavoidable on a grand scale. The problem is how to mitigate the craziness entailed by them, and the only way to do this once such bizarre possibilities are conceded is to render them improbable on a local scale. Thus, in the case of Rubinstein, there are worlds where someone named Artur Rubinstein is a world famous pianist and does not know the first thing about music. But it is vastly more probable that in worlds where someone named Artur Rubinstein is a world famous pianist, that person is a consummate musician. What's more, induction tells us that ours is such a world.

But can induction really tell us that? How do we know that we are not in one of those bizarre worlds where things that we ordinarily attribute to design actually happen by chance? Consider further the case of Artur Rubinstein. Imagine it is January 1971 and you are at Orchestra Hall in Chicago listening to Rubinstein perform. As you listen to him perform Liszt's Hungarian Rhapsody no. 2 in C-sharp Minor, you think to yourself, *I know the man I'm listening to right now is a wonderful musician. But there's an outside possibility that he doesn't know the first thing about music and is just banging away at the piano haphazardly.* The fact that Liszt's Hungarian Rhapsody is cascading from Rubinstein's fingers would thus merely be a happy accident.

What's more, the "fact" that he has done this at many other places, that he has often spoken eloquently about music and piano, and that it is well documented that he spent many years seemingly developing his skills— these are also all part of the happy coincidence afforded by an infinity of universes. He's just a lucky imbecile banging away at a keyboard in just such a way as to coincidentally give the appearance of a man dramatically progressing in his skills, babbling strings of sounds that just happen to make for astute music commentary. What a fun coincidence! Of course the idea that Rubinstein was just banging away at a keyboard and getting lucky is absurd. But if I take seriously the existence of infinite other worlds, then there is some counterpart to me pondering these very same thoughts, only this time listening to the performance of someone named Artur Rubinstein who is a complete musical ignoramus. How, then, do I know that I'm not that counterpart?

Indeed, how do you know that you are listening to Artur Rubinstein the musical genius and not Artur Rubinstein the lucky poseur? To answer this question, let us ask a prior question: What led you to think that the man called Rubinstein performing in Orchestra Hall was a consummate musician? His reputation, his formal attire and the famous concert hall are certainly giveaways, but they are neither necessary nor sufficient. Even so, a necessary condition for recognizing Rubinstein's musical skill (i.e., design) is that he was following a prespecified concert program, in this instance that he was playing Liszt's Hungarian Rhapsody no. 2 in C-sharp Minor note for note (or largely so—Rubinstein made lots of mistakes). In other words, you recognized that Rubinstein's performance exhibited specified complexity. Moreover, the degree of specified complexity exhibited enabled you to assess just how improbable it was that someone named Rubinstein was playing the Hungarian Rhapsody with apparent proficiency but did not have a clue about music. Granted, you may have lacked the probabilistic vocabulary to describe the performance in these terms, but the recognition of specified complexity was there nonetheless, and without that recognition there would have been no way to attribute Rubinstein's playing to design rather than chance.

We use specified complexity to eliminate bizarre possibilities in which chance is made to account for things that we would ordinarily attribute to design. What's more, we use specified complexity to assess the improbability of those bizarre possibilities and to justify eliminating their chance occurrence. That being the case (and it certainly is the case for human artifacts), on what basis could we attribute natural phenomena that exhibit specified complexity to chance? Note that we are not just talking about an analogy here (e.g., that artifacts are similar biological systems). Rather, we are talking about an *isomorphism*: the specified complexity in artifacts is identical with the specified complexity in biological systems.

On what basis, then, could we attribute natural phenomena that exhibit specified complexity to chance? Let's be clear that inflating probabilistic resources does not just diminish a universal probability bound and make it harder to attribute design; inflating probabilistic resources is not a matter of replacing one universal probability bound with another that is more stringent. Inflating probabilistic resources eliminates universal probability bounds entirely: the moment one posits unlimited probabilistic resources, anything of nonzero probability is sure to happen somewhere at sometime. (This follows from what probabilists call the *strong law of large numbers*.)

It seems, however, that in practical life we do allow for probability bounds to assess improbability and therewith specified complexity. A sentence or two verbatim repeated by another author can be enough to elicit the charge of plagiarism. It could happen by chance, and given unlimited probabilistic resources there are patches of reality where it did happen by chance. But we don't buy it—at least not for our patch of reality. In practical life we tend not to be very conservative in setting probability bounds. In other words, in practical life, we don't demand that something be anywhere near as unlikely as the universal probability bound of 1 in 10^{150} before we rule out chance and infer design.

The difficulty confronting unlimited probabilistic resources can now be put quite simply: there is no principled way to discriminate between using unlimited probabilistic resources to retain chance and using specified complexity to eliminate chance. You can have one or the other, but you cannot have both. And the fact is, we already use specified complexity to eliminate chance. Let me stress that there is no principled way to make the discrimination. Of course people still make it. For instance, naturalistic scientists often invoke naturalism as a philosophical presupposition and unlimited probabilistic resources as a tool for retaining chance when designers unacceptable to naturalism are implicated (e.g., God). Then they will turn around and use specified complexity to eliminate chance when designers acceptable to naturalism are implicated (e.g., Francis Crick's space aliens who seed the earth with life as in his theory of directed panspermia). Thus, for artificial objects that exhibit specified complexity and for which an embodied intelligence could plausibly have been involved, they would attribute design; but for natural objects that exhibit specified complexity and for which no embodied intelligence could plausibly have been involved, they would invoke unlimited probabilistic resources and thus attribute chance (or perhaps simply plead ignorance). Why? Not because it's logical. Indeed, the maneuver is entirely arbitrary. The problem of unlimited probabilistic resources, by raising the question of specified complexity and design inferences, throws naturalism itself into question, and it does no good to invoke naturalism to resolve the problem.

We are now in a position to see why a designer outside the known universe could in principle be supported by independent evidence whereas an infinite universe never can. We already have experience of human and animal intelligences generating specified complexity. If we should ever discover evidence of extraterrestrial intelligence, a necessary feature of that evidence would be specified complexity. Thus, when we find specified complexity in nature which no embodied, reified or evolved intelligence could plausibly have placed there, it is a straightforward inference to conclude that some unembodied intelligence must have been involved. Granted, this raises the question of how such an intelligence could coherently interact with the physical world. (More on this in chapters twenty and twenty-six.) But to deny this inference merely because of a prior commitment to naturalism is not defensible. There is no principled way to distinguish between using specified complexity to eliminate chance in one instance and then in another instance invoking unlimited probabilistic resources to render chance plausible.

Design, therefore, as signified by specified complexity, allows for the possibility of independent evidence; an infinite universe with unlimited probabilistic resources does not. Specified complexity can be a point of contact between the known universe, which is finite, and an intelligence outside it. Designers within the known universe already generate specified complexity, and a designer outside could potentially do the same. That is what allows for independent evidence to support unembodied designers. Provided nature supplies us with instances of specified complexity that cannot reasonably be attributed to any embodied intelligence, the inference to an unembodied intelligence becomes compelling and any instances of specified complexity used to support that inference can rightly be regarded as independent evidence.

It appears, then, that we are back to our own known little universe, with its very limited number of probabilistic resources but also its increased possibilities for detecting design. This is one instance where less is more, where having fewer probabilistic resources and a smaller universe opens possibilities for knowledge and discovery that would otherwise be closed. Limited probabilistic resources enrich our knowledge of the world by enabling us to detect design where otherwise it would elude us. At the same time, limited probabilistic resources protect us from the unwarranted confidence in natural causes that unlimited probabilistic resources invariably seem to engender. In short, limited probabilistic resources eliminate the chance of the gaps.

PART 3

INFORMATION

16

INFORMATION AND MATTER

*What is the difference between information
and matter, and what role does each play
in the theory of intelligent design?*

IMAGINE YOU ARE AN INTERIOR DECORATOR. Bill and Melinda French
Gates hire you to decorate their mansion. You decide to put a big marble
bust of the composer Ludwig van Beethoven in the music room, so you
contract with a promising if eccentric young sculptor to make the bust and
deliver it to the mansion. The next day, he drops by the mansion and tells
you he's finished. Naturally, you're skeptical, but he rolls an imposing
crate into the music room and with some fanfare removes the "sculpture."
You find yourself staring at a big marble cube. Shocked, you ask the sculptor where the bust of Beethoven is. "The bust is there all right," he says,
handing you his bill. "You just have to scoot aside the excess marble."
When you protest, he grows red faced and yells, "I defy you to find a single, solitary molecule of Beethoven's bust that isn't in that block of marble!
Cross me and I'll sue you and your billionaire boss Mr. Gates right into the
poor house!"

The quarrel continues to escalate, each of you growing more and more
red-faced until who should wander into the room but Bill Gates himself.
He calms the two of you and gets first your side of the argument and then
the sculptor's. "Tell you what," Gates says after he's heard the sculptor
out. "I'm so impressed with this sculpture that instead of paying you what
we agreed, I'll trade you an advance copy of the next generation of Microsoft *Windows*." Here Gates produces an unmarked compact disc. "This
is it, and you're free to sell the operating system on the black market to
whomever you wish as often as you wish." "Are you crazy?" you shout,
forgetting yourself. "That's worth billions!" "Deal!" the sculptor shouts,
snatches the disc from Gates's hand, and rushes to a nearby computer. After some pointing and clicking, the sculptor turns on Gates. "The disc is

blank! Give me the operating system!" "Oh, but friend," Gates says, resting an avuncular hand on the young sculptor's shoulder, "I defy you to find a single, solitary molecule of the operating system that isn't on that disc. You just have to scoot aside the excess polycarbonate and there it is."

This story illustrates the difference between information and matter. Matter is raw stuff that can take any number of shapes. Information is what gives shape to matter, fixing one shape to the exclusion of others. Both the words *matter* and *information* derive from Latin. *Matter* (from the Latin noun *materia*) initially referred to the raw timber used in building houses. Later it came to mean any raw stuff or material with the potential to assume different shapes, forms or arrangements. *Information* (from the Latin verb *informare*) literally means to give form or shape to something. Unlike passive or inert matter, which needs to be acted upon, information is active. Information acts on matter to give it its form, shape, arrangement or structure. (Note that I'm using these terms loosely and interchangeably. Aristotle would distinguish form, in the sense of substantial form or essence, from mere shape or arrangement. It's enough for my purposes here, however, that shape or arrangement be correlated with form in Aristotle's sense. Thus for marble to express the form of Beethoven's likeness, it must be shaped or arranged in very particular ways.)

As an interior decorator, you were paying the sculptor to "inform" a slab of marble—to take an unformed slab of marble and give it the form or shape of Beethoven. For the sculptor to tell you that the cube of marble contains the promised bust of Beethoven (you have only to remove some excess marble) is therefore totally unacceptable. That's what you were paying the sculptor to do. Yes, the marble cube has the potential to become a bust of Beethoven. But it also has the potential to take on countless other shapes. It was the sculptor's job to give the marble the shape you requested.

The relation between matter, with its potential to assume any possible shape, and information, with its restriction of possibilities to a narrow range of shapes, is fundamental to our understanding of the world. Certainly, this relation holds for all human artifacts. This is true not only for human artifacts composed of physical stuff (like marble busts of Beethoven), but also for human artifacts composed of more abstract stuff (like poetry and mathematics). Indeed, the raw material for many human inventions consists not of physical stuff but of abstract stuff like alphabetic characters, musical notes and numbers. For instance, the raw material for a Shakespearean sonnet consists of the twenty-six letters of the alphabet. Just as a bust of Beethoven is only potential in a slab of marble, so a Shakespearean sonnet is only potential in those twenty-six letters. It takes a sculptor to actualize the bust of Beethoven, and it takes a Shakespeare to arrange those twenty-six letters appropriately so that one of his sonnets emerges.

The relation between matter and information that we are describing here is old and was understood by the ancient Greeks, especially the Stoics. What's more, nothing said so far about the relation between matter and information is especially controversial. The world consists of a lot of raw material waiting to be suitably arranged. There's matter, passive or inert stuff waiting to be arranged, and there's information, an active principle or agency that does the arranging. This is a perfectly straightforward and useful way of carving up experience and making sense of the world. Much of our knowledge of the world depends on understanding the relation between matter and information.

Nonetheless, the relation between matter and information does become controversial once we add another dimension to it. That happens when we place matter and information in combination with design and nature.

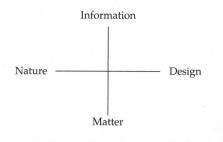

Figure 3

So far the examples of information that we've considered have focused on the activity of a designing intelligence (a sculptor or writer) informing or giving shape to certain raw materials (a slab of marble or letters of the alphabet). But designing intelligences are not the only agents capable of structuring matter and thereby conferring information. Nature, too, is capable of structuring matter and conferring information.

Consider the difference between raw pieces of wood and an acorn. Raw pieces of wood do not have the power to assemble themselves into a ship. For raw pieces of wood to form a ship requires a designer to draw up a blueprint for a ship, then take the pieces of wood and, in line with the blueprint, fashion them into a ship. But where is the designer that causes an acorn to develop into a full-grown oak tree? There isn't any. The acorn has within itself the power to transform itself into an oak tree.

Nature and design therefore represent two different ways of producing information. Nature produces information, as it were, internally. The acorn assumes the shape it does through powers internal to it: the acorn is a seed programmed to produce an oak tree. On the other hand, a ship assumes the shape it does through powers external to it: a designing intelligence imposes a suitable structure on pieces of wood to form a ship.

Not only did the ancient Greeks know about the distinction between information and matter, but they also knew about the distinction between design and nature. For Aristotle, for instance, design consisted in capacities external to an object for bringing about its form with outside help. On the other hand, nature consisted in capacities internal to an object for transforming itself without outside help. Thus in book twelve of the *Metaphysics*, Aristotle wrote, "[Design] is a principle of movement in something other than the thing moved; nature is a principle in the thing itself." In book two of the *Physics* Aristotle referred to design as completing "what nature cannot bring to a finish." (Note that Thomas Aquinas took this idea and sacramentalized it into grace completing nature.)

The Greek word here translated "design" is *techne*, from which we get our word *technology*. In translations of Aristotle's work, the English word most commonly used to translate *techne* is "art" (in the sense of "artifact"). *Design, art* and *techne* are synonyms. The essential idea behind these terms is that information is conferred on an object from outside the object and that the material constituting the object, apart from that outside information, does not have the power to assume the form it does. For instance, raw pieces of wood do not by themselves have the power to form a ship.

This contrasts with nature, which does have the power within itself to express information. Thus in book two of the *Physics* Aristotle wrote, "If the ship-building art were in the wood, it would produce the same results by nature." In other words, if raw pieces of wood had the capacity to form ships, we would say that ships come about by nature. The Greek word here translated *nature* is *physis*, from which we get our word *physics*. The Indo-European root meaning behind *physis* is growth and development. Nature produces information not by imposing it from outside but by growing or developing informationally rich structures from within. Consider again the acorn. Unlike wood, which needs to be fashioned by a designer to form a ship, acorns produce oak trees naturally; the acorn simply needs a suitable environment in which to grow.

The central issue in the debate over intelligent design and biological evolution can therefore be stated as follows: is nature complete in the sense of possessing all the resources needed to bring about the information-rich biological structures we see around us, or does nature also require some contribution of design to bring about those structures? Aristotle claimed that the art of shipbuilding is not in the wood that constitutes the ship. We've seen that the art of composing sonnets is not in the letters of the alphabet. Likewise, the art of making statues is not in the stone out of which statues are made. Each of these cases requires a designer. So too, the theory of intelligent design contends that the art of building life is not in the physical stuff that constitutes life but requires a designer.

17

INFORMATION THEORY

How does the mathematical theory of information relate to intelligent design and, specifically, to intelligent design's criterion for detecting design, namely, specified complexity?

ORDINARILY, WHEN WE THINK OF INFORMATION, we think of meaningful statements that we communicate to each other. The vehicle of communication here is language, and the information is the meaning communicated by some utterance or linguistic expression. This picture of information diverges sharply from the picture of information associated with the mathematical theory of information. The ordinary picture of information focuses on meaning and treats the linguistic vehicle by which that meaning is transmitted as secondary. The mathematical picture of information, by contrast, focuses exclusively on the vehicle and ignores the meaning entirely.

Consider a spy who needs to determine the intentions of an enemy—whether that enemy intends to go to war or preserve the peace. The spy agrees with headquarters about which signal will indicate war and which signal will indicate peace. Let's imagine that the spy will send headquarters a radio transmission and that each transmission takes the form of a bit string (i.e., a sequence of zeroes and ones). The spy and headquarters might therefore agree that 0 means war and 1 means peace. But because noise along the communication channel might flip a 0 to a 1 and vice versa, it might be good to have some redundancy in the transmission. Thus the spy and headquarters might agree that 000 represents war and 111 peace and that anything else will be regarded as a garbled transmission. Or perhaps they will agree to let the spy communicate via Morse code in plain English whether the enemy plans to go to war or maintain peace.

This example illustrates how information, in the sense of meaning, can remain constant whereas the vehicle for representing and transmitting this information can vary. In ordinary life we are concerned with meaning. If we are at headquarters, we want to know whether we're going to war or stay at peace. Yet from the vantage of mathematical information theory, the only thing important here is the mathematical properties of the linguistic expressions we use to represent the meaning. If we represent war with 000 as opposed to 0, we require three times as many bits to represent war, and so from the vantage of mathematical information theory we are using three times as much information. The information content of 000 is three bits whereas that of 0 is just one bit.

Claude Shannon invented the mathematical theory of information shortly after World War II. The inspiration for his theory derived from his work during the war on cryptography. In cryptography, meaningful messages get encrypted to prevent an enemy from reading one's mail. The important thing in cryptography is to have a secure encryption-decryption scheme: to be able to code messages efficiently as character strings from some alphabet and then to be able to move those character strings efficiently across communication channels.

The actual meaning of a character string therefore takes second seat in the mathematical theory of information. Think of the mathematical theory of information as an Internet service provider. The Internet service provider is not concerned with the meaning of your e-mail messages or with the product you're trying to sell on your website. What it's concerned about is that the character strings you use to convey meaning in your e-mails or on your website is faithfully stored and transmitted. That's what the mathematical theory of information is all about. Specifically, it is about quantifying the information in such character strings, characterizing the statistical properties of such strings when they are sent across a noisy communication channel (noise typically is represented as a stochastic process that disrupts the strings in statistically well-defined ways), preserving the strings despite the presence of noise (i.e., the theory of error-correcting codes), compressing the strings to improve efficiency and transforming the strings into other strings to maintain their security (i.e., cryptography).

Although Shannon's theory started out as a syntactic theory concerned with character strings based on a fixed alphabet, it quickly became a statistical theory. Characters from an alphabet will often have different probabilities of occurrence. (For instance, the letters from our ordinary alphabet occur with widely varying frequencies: in English the letter *e* occurs roughly 12 percent of the time, the letter *q* less than 1 percent of the time;

what's more, u follows q 100 percent of the time.) These probabilities in turn determine how much information any given string can convey. In general, the quantity of information contained in a character string corresponds to the improbability of that character string. Thus, the more improbable the string, the more information it contains.

To see why this should be the case, consider the claim "it's raining outside." This claim will be more informative (now in a loose semantic sense) depending on how improbable it is. If it refers to weather in the Sahara Desert during the summer, when the chance of rain is very low, then this claim will be both highly improbable and highly informative: it's telling you something you wouldn't otherwise have guessed. But if this claim refers to weather in Seattle during the spring, when the chance of rain is very high, then it will be both probable and uninformative: it's telling you something you could easily have guessed. The mathematical theory of information models this feature of our ordinary understanding of information, making high-probability claims have low information content and low-probability (high-improbability) claims have high information content.

Given this characterization of high and low information in terms of probability, there's no reason to confine the mathematical theory of information to character strings. Indeed, any reference class of possibilities over which there is a probability distribution is fair game for the mathematical theory of information. For information to be generated, therefore, means identifying one possibility and ruling out the rest. The more possibilities get ruled out and, correspondingly, the more improbable the possibility that actually obtains, the greater the information generated. To rule out no possibilities is to assert a tautology and provide no information. "It's raining or it's not raining" is true but totally uninformative. On the other hand, "it's raining" is informative because it rules out "it's not raining." Moreover, "it's raining" is informative to the degree that this claim is improbable. (This claim is therefore going to be more informative in the Sahara Desert than in Seattle.)

To generate information is therefore to rule out possibilities. Moreover, the amount of information generated here corresponds to the probability of the possibility (or range of possibilities) that wasn't ruled out. But who or what rules out possibilities? In practice, there are two sources of information: intelligent agency and physical processes. This is not to say that these sources of information are mutually exclusive; human beings, for instance, are both intelligent agents and physical systems. Nor is this to say that these sources of information exhaust all logically possible sources of information; it is conceivable that there could be nonphysical random processes that generate information.

Although physical processes that are not also intelligent agents can generate information, there is a sense in which information, whatever its source, is irreducibly conceptual and thus presupposes intelligent agency. That is because the very reference class of possibilities that sets the backdrop for the generation of information must invariably be delineated by an intelligent agent. Thus information, whatever else we might want to say about it, can never be entirely mind-independent or concept-free.

Nevertheless, once an intelligent agent identifies a reference class of possibilities, it is a separate question whether information generated from that reference class results from an intelligent agent or a physical process. An intelligent agent may explicitly identify a pattern within the reference class of possibilities and thereby generate information. Alternatively, a physical process can produce an event, represented as a possibility within the reference class of possibilities, and thereby generate information. Let us refer to the former type of information as *agent-induced* or *conceptual information* and to the latter as *event-induced* or *physical information*.

Now, what happens when conceptual information and physical information coincide? For instance, what happens if, as a conceptual act, SETI researchers identify a sequence of prime numbers and then, lo and behold, as in the movie *Contact*, that very sequence is transmitted, as a physical event, to the radio telescopes that these same SETI researchers are monitoring? It's precisely such a coincidence that constitutes specified complexity.

But note that within Shannon's theory of information, such a coincidence plays no role. Shannon's theory is simply concerned with generating information from a reference class of possibilities. It is immaterial to Shannon's theory whether the information generated is agent-induced or event-induced. Specified complexity, by contrast, requires a dual ruling out of possibilities, one by an intelligent agent who identifies a pattern and one by physical processes that induce an event. Provided these coincide, the probability is small and the pattern can be identified independently of the event, we say the event exhibits specified complexity.

Specified complexity (or complex specified information, as it's also called) is therefore a souped-up form of information. To be sure, specified complexity is consistent with the basic idea behind information, which is the reduction or ruling out of possibilities from a reference class of possibilities. But whereas the traditional understanding of information is unary, conceiving of information as a single reduction of possibilities, specified complexity is a binary form of information. Specified complexity depends on a dual reduction of possibilities, a conceptual reduction (i.e., conceptual information) combined with a physical reduction (i.e., physical information). Moreover, these dual reductions must be coordinated so that the physical information matches the pattern set by the conceptual information.

This information-theoretic approach to specified complexity is exactly equivalent to the design-theoretic approach to specified complexity associated with the design inference, the complexity-specification criterion and the Explanatory Filter described earlier (see part two). Here, then, is the connection between intelligent design and information theory: detecting design by means of the complexity-specification criterion is equivalent to identifying complex specified information.

18

BIOLOGY'S INFORMATION PROBLEM

What is biology's information problem, and how do biologists attempt to resolve it?

IN A WIDELY CITED SPEECH, NOBEL LAUREATE David Baltimore remarked, "Modern biology is a science of information." Manfred Eigen, Bernd-Olaf Küppers, John Maynard Smith and many other biologists have likewise identified information as biology's central problem. For matter to be alive, it must be suitably structured. A living organism is not a mere lump of matter. Life is special, and what makes life special is the arrangement of its matter into very specific forms. In other words, what makes life special is information. Where did the information necessary for life come from? This question cannot be avoided. Life has not always existed. There was a time in the history of the universe when all matter was lifeless. And then life appeared—on earth and perhaps elsewhere.

The appearance of life constitutes a revolution in the history of matter. A vast gulf separates the organic from the inorganic world, and that gulf is properly characterized in terms of information. The matter in the dirt under your feet and the matter that makes up your body is the same. Nevertheless, the arrangement of that matter—the information—vastly differs in these two cases. Biology's information problem is therefore to determine whether (and if so, how) purely natural forces are able to bridge the gulf between the organic and inorganic worlds as well as the gulfs between different levels of complexity within the organic world. Conversely, biology's information problem is to determine whether (and if so, how) design is needed to complement purely natural forces in the origin and subsequent development of life.

Not all biological structures or arrangements are equally relevant to deciding whether life is designed. For instance, Darwin's mechanism of natural selection acting on random variation is responsible for certain features of biological systems. Antibiotic resistance in bacteria and insecticide resistance in insects can be readily accounted for in terms of the Darwinian mechanism. But can that mechanism or any other purely natural mechanism account for how we got bacteria and insects in the first place? And if not, what is it about the information exhibited by such systems that conclusively tells us they could not have been formed by purely natural means?

To say that something formed by purely natural means is to say that it resulted from necessity, chance or a combination of these. To show that some biological system is designed therefore requires, at a minimum, showing that it could not reasonably have formed by these means. This in turn requires ruling out necessity, chance and their combination as sufficient to account for the biological system in question. This is not to say that natural forces were uninvolved. For instance, just as a rusted old automobile shows the effects of both design (engineering) and natural forces (weathering and corrosion), so too biological systems can exhibit the effects of design and natural forces.

Let us consider chance, necessity and their combination a bit more closely. Consider first a biological structure that results from necessity. It would have to form as reliably as water freezes when its temperature is suitably lowered. As a consequence, there would be no need to explain that structure in terms of design; a lawlike regularity of nature would readily account for it. True, one might want to attribute design to the necessitarian process responsible for some biological structure, but that's a different question. Attributing design in that case would be like attributing design to water's propensity to freeze below a certain temperature. Perhaps water is designed that way. But such necessitarian processes are as readily deemed brute facts of nature as artifacts of design. The problem is not that explaining some object as the result of a necessitarian process precludes design, but that it can never decisively implicate design.

Next, consider chance. If some biological structure is reasonably likely, there is no reason to invoke design. Chance—whether pure chance or chance combined with necessity—handles that case without difficulty. Likewise, if some biological structure is incredibly unlikely but fails to be specified, there is again no reason to invoke design in its explanation. For example, the precise configuration of moles and skin imperfections on your back are highly unlikely, but there's nothing in that pattern that specifies it. It is an arbitrarily rather than independently given pattern.

What we've just done is argue that specified complexity is a necessary condition for detecting biological design. Thus, to rule out that a biological system is due to purely natural forces, we have to show that it is contingent, complex and specified. In that case we say it exhibits *specified complexity*. In earlier chapters I argued that specified complexity is a reliable criterion for detecting design. Accordingly, specified complexity is not merely a necessary but also a sufficient condition for detecting design.

Now, as we saw in chapter seventeen, specified complexity is also a type of information. Indeed, it is the only type of information that can reliably detect design. For unless something is contingent, complex and specified, it is readily referred to necessity, chance or their combination. Without specified complexity, all biological explanation reduces to natural mechanisms (i.e., mechanisms that invoke only necessity, chance or their combination). Specified complexity is therefore the key to resolving biology's information problem.

Consequently, if specified complexity is exhibited in actual biological systems, we are justified attributing such systems to design. That's not to say that every aspect of such systems is designed. (Some aspects may be due to purely natural forces.) But just as human artifacts are legitimately attributed to design even if they show signs of wear due to natural forces, so biological systems that exhibit specified complexity are legitimately attributed to design even though purely natural forces have also modified them over time. Organisms exhibit design, but they also exhibit the effects of history, and purely natural forces have operated in history.

One of the main tasks facing intelligent design as a scientific research program is to trace the specified complexity of biological systems and thereby demonstrate points at which design decisively enters biology. Specified complexity is a reliable empirical marker of intelligence in the same way that fingerprints are a reliable empirical marker of a person's presence. Specified complexity therefore reliably detects design. Once an object exhibits specified complexity, we can be confident that it is designed. To establish the specified complexity of an object, however, requires showing two things: that it is specified and that it is complex or improbable.

In practice, specification is never a problem. Many biological structures are specified in virtue of independently given functional requirements that they satisfy. Consider, for instance, the bacterial flagellum, a bidirectional motor-driven propeller that moves certain bacteria through their watery environment. One way to see that the bacterial flagellum is specified is that humans developed such motor-driven propellers well before they figured out that the flagellum was such a machine. (Such propellers were in common use with ships in the nineteenth century, yet the function of the flagellum was not discovered until the early twentieth century.) This is not to say that for the biological function of a system to constitute a specification, humans must have independently invented a system that performs the same function. Nevertheless, independent invention makes the specification all the more stark.

At any rate, no biologist questions whether the functional systems that arise in biology are specified. At issue always is whether the Darwinian mechanism or some other purely natural mechanism can overcome the vast improbabilities that at first blush seem to arise for such systems. Such a mechanism would break a seemingly vast improbability into a series of more manageable probabilities. The task of such a mechanism is therefore to render probable what otherwise seems highly improbable.

Mechanistically inclined biologists, in trying to explain specified complexity, therefore always end up explaining away the complexity or improbability associated with specified biological systems. To attribute complexity to something, it must be highly improbable with respect to all natural mechanisms currently known. Consequently, for a natural mechanism to come along and explain something that previously was regarded as complex means that the item in question is in fact no longer complex with respect to the newly found mechanism.

There are several strategies for explaining away complexity mechanistically. Since these strategies come up repeatedly in discussions over intelligent design, it will help to lay them out here.

Strategy 1: Spontaneous generation. To say that a biological structure arose by spontaneous generation is to say that it arose by itself all at once. One moment it wasn't there, the next moment it was. To invoke spontaneous generation is to appeal to pure chance. The actual term *spontaneous generation* used to be quite common. Just 150 years ago it was widely thought that flies and mice could form spontaneously from rotting meat and dirty rags. The spontaneous generation of complex multicelled organisms like flies and mice is no longer taken seriously in the biological community. Nonetheless, origins-of-life researchers do take seriously the spontaneous generation of less complex biological structures, like simple self-replicating molecules. If they are simple enough, then they may not be too improbable and therefore could reasonably be expected to form by pure chance.

Strategy 2: Divide and conquer. Biologists who reject design have no choice but to explain the emergence of highly complex biological systems in terms of the evolution of simpler systems. Given some highly complex biological system, there is no way that it could form by spontaneous generation—it's just too improbable. As a consequence, there has to be some precursor system that is simpler and therefore less improbable than the system in question, and from which this system might with reasonable probability have evolved through a blind evolutionary process. If the precursor system is too improbable to have formed by spontaneous generation, then that system too will require a precursor, which in turn may require a precursor, and so on until we get to a system that can form by spontaneous generation. The Darwinian mechanism of random variation and natural selection is biology's premier divide-and-conquer strategy. Accordingly, each generation in an organism's evolutionary history constitutes a round of random variation and natural selection and therefore a baby step in the organism's evolutionary path. The Darwinian mechanism is a trial-and-error mechanism, with natural selection providing the trial and random variation providing the error. As with all trial-and-error mechanisms, the Darwinian mechanism hinges on slow, gradual improvements. Insofar as it succeeds at all, the Darwinian mechanism succeeds by numerous divisions and numerous small conquests.

Strategy 3: Self-organization. Darwinism makes chance, in the form of random variation, a creative force in biology. At the same time, the Darwinian mechanism does not allow chance to run unchecked. It introduces natural selection to constrain random variation, picking out those random variations that confer biological fitness. It follows that chance is the source of creativity of the Darwinian mechanism. By contrast, self-organization makes necessity rather than chance the source of biological creativity. If the image of the Darwinian mechanism is a series of baby-steps up a high mountain (cf. Richard Dawkins's *Climbing Mount Improbable*), then the image of self-organization is a whirlpool that orders a fluid and inescapably carries it downward. Chance still operates with a whirlpool. For instance, how the whirlpool oscillates around its center will be determined by chance. But the global self-organizational "whirling" behavior of the whirlpool is a matter of necessity and not chance. According to the self-organizational approach to biological complexity, just as water under the right conditions produces a whirlpool, so the laws of physics and chemistry build into matter self-organizational properties that, under the right conditions, produce complex biological structures.

Strategy 4: Pass the buck. Suppose there's a hole you have to fill. One way to do it is to dig another hole, take the dirt from it and fill the first hole.

But now you've got another hole to fill. Many attempts to deal with specified complexity adopt this approach, filling one hole by digging another. In other words, they pass the buck. Perhaps the most celebrated instance of passing the buck is Francis Crick's theory of "directed panspermia." Panspermia refers to the seeding of life from outer space. In some accounts of panspermia, microbes hitch a ride on asteroids, make it to planets like earth and thereby introduce life to planets (life that then goes on to evolve). Panspermia theories attempt to get around the problem of life's origin (though not its evolution). In Crick's theory of directed panspermia, microbes don't hitch a ride on an asteroid but are instead intentionally carried by intelligent space aliens who deliberately place them on planets like earth. Crick holds this view because he sees life as too fragile to survive on asteroids whizzing through space (spaceships are much safer) and also because he sees life as too improbable to have arisen on earth. Crick therefore passes the buck to some unknown place in the universe where life was more likely to have arisen by purely natural forces. Life on earth is thereby "explained," but life at some unknown location in the universe now becomes a mystery. Passing the buck is not a solution. It attempts to counter the challenge of specified complexity not by actually refuting it but by gesturing at the possibility of a naturalistic solution elsewhere. It postpones rather than solves.

These are the four main strategies that biologists use to circumvent the challenge of specified complexity and thereby attempt to resolve biology's information problem. Sometimes these strategies are used individually, sometimes in combination. Crick, for instance, will use "passing the buck" to get life started and then "divide and conquer" to explain life's subsequent evolution. Stuart Kauffman will use "spontaneous generation" to get a simple replicator going, use "divide and conquer" to account for the day-to-day operations of evolution and then invoke "self-organization" to get evolution over certain humps that "divide and conquer" cannot handle. All such strategies have uniformly failed to make headway with the specified complexity of biological systems like the bacterial flagellum.

19

INFORMATION EX NIHILO

Is nature complete in the sense of possessing all the capacities needed to bring about the information-rich structures that we see in the world and especially in biology? Or are there informational aspects of the world that nature alone cannot bridge but require the guidance of an intelligence?

THIS QUESTION ASKS WHETHER NATURE, conceived in nonteleological or purely naturalistic terms, can by itself bring about all its various productions. In the context of biology, we might put the question this way: Does nature have what it takes to produce life, and if not, what else does it need? There's no question that nature supplies the raw materials for life. But that's a different question. The question is whether those raw materials contain within themselves the power or capacity to produce life. And if not, what has the power or capacity in conjunction with nature to produce life?

First off, let's be clear that the question we are asking here is not nearly as simple as asking whether the letters of the alphabet have the power to produce a meaningful text. Clearly, the letters of the alphabet by themselves do not have that power or capacity. Karl Marx used to joke that the twenty-six letters of the alphabet were the soldiers with which he would conquer the world. Some would say he almost succeeded. But note that the soldiers required Marx's direction (or generalship, to continue his military metaphor). But by themselves, the letters of the alphabet are inert. They do not, for instance, have what it takes to produce Marx's *Das Kapital*. Marx and the letters of the alphabet, however, do. But that raises the question how Marx himself came about.

From a Darwinian perspective, any designing intelligence like Marx results from a long and blind evolutionary process. Nature, unassisted and unguided by any intelligence, starts off from a lifeless earth and, over the course of natural history, produces life forms that eventually evolve into human beings like Karl Marx, who then write economic treatises like *Das Kapital*. Within Darwinism, only natural forces, like natural selection and random variation, control the evolutionary process. Designing intelligences may evolve out of that process but have no place in guiding or controlling it.

But how can we determine whether nature has what it takes to produce life? Don't let the sheer commonness of life fool you. We look around and see life everywhere. But there was a time when the earth housed no multicelled organisms like us. Before that, the earth housed no life at all, not even single-celled forms. And earlier still, there was no earth at all, no sun or moon or other planets. Indeed, if physicists are right, there was a time when there were no stars or galaxies but only elementary particles like quarks densely packed at incredibly hot temperatures. That would coincide with the moment just after the big bang.

Suppose we go back to that moment. Given the history of the universe since then, we could say—in retrospect—that all the possibilities for complex living forms like us were in some sense present at that earlier moment in time (much as many possible statues are in some sense present in a block of marble). From that early state of the universe, galaxies and stars eventually formed, then planet earth formed, followed by single-celled life forms and finally life forms as complicated as us. But that still doesn't tell us how we got here or whether nature had sufficient creative power to produce us apart from design.

In *Genes, Genesis and God*, philosopher Holmes Rolston puts his finger on the problem. He notes that humans are not invisibly present in primitive single-celled organisms in the same way that an oak tree is secretly present in an acorn. The oak tree unfolds in a lawlike or programmatic way from an acorn. But the same cannot be said for the grand evolutionary story that places single-celled organisms at one end and humans at the other. There's no sense in which human beings or even multicelled organisms are latent in single-celled organisms, much less in nonliving chemicals. According to Rolston, to claim that life is somehow already present and lurking in nonliving chemicals, or that complex biological systems are already present and lurking in simple biological systems, is "an act of speculative faith."

In science it is not enough merely to assert a claim; one must also back it up. It is not enough, therefore, merely to assert that nature possesses the causal powers necessary to produce living forms. Rather, one must demonstrate that nature actually does possess such causal powers. What's more, we don't have the luxury, as did Aristotle and many ancient philosophers, of thinking that life and the universe have always been here. We now know that the earth has not been here forever and that the early earth was both uninhabited and uninhabitable—a barren, searing and tempestuous cauldron.

Yet somehow, from fairly simple inorganic compounds on the early lifeless earth, life forms requiring incredibly complex biomacromolecules emerged. How did that happen? How could it have happened? Indeed, what made it happen? Now we can conjecture that blind natural forces all by themselves made it happen. But if so, how can we know it? And if not, how can we know that? According to the theory of intelligent design, the specified complexity exhibited in living forms convincingly demonstrates that blind natural forces could not by themselves have produced those forms but that their emergence also required the contribution of a designing intelligence.

The design found in nature therefore demonstrates that nature is incomplete. In other words, nature exhibits design that nature is unable to account for. What's more, since the design in nature is identified through specified complexity, and since specified complexity is a form of information and since this form of information is beyond the capacity of nature, it follows that specified complexity and the design it signifies is information ex nihilo. That is, it's information that cannot be derived from natural forces acting on preexisting stuff. Indeed, to attribute the specified complexity in biological systems to natural forces is like saying that Scrabble pieces have the power to arrange themselves into meaningful sentences. The absurdity is equally palpable in both cases. Only in evolutionary biology the absurdity has been repeated so often that we no longer recognize it.

Nature's incompleteness in accounting for design does not mean that design in nature is a miracle or that it requires supernatural intervention. With the rise of modern science, miracles have come to refer to a supernatural intervention that violates or suspends or overrides natural laws. To attribute something to a miracle is therefore to say that a natural cause was all set to make one thing happen but instead another thing happened (the miracle). Design does not require this sort of counterfactual substitution (where what didn't happen—the counterfactual—was supposed to have happened if natural laws had not been interrupted).

When humans, for instance, act as intelligent agents, there is no reason to think that any natural law is broken. Likewise, should a designer act to bring about a bacterial flagellum, there is no reason to suppose that this designer did not act consistently with natural laws. It is, for instance, a logical possibility that the design in the bacterial flagellum was front-loaded into the universe at the big bang and subsequently expressed itself in the course of natural history as a miniature motor-driven propeller motor on the back of *E. Coli*. Whether this is what actually happened is another question, but such a possibility involves no contradiction of natural laws and gets around the usual charge of miracles.

Nonetheless, even though intelligent design requires no contradiction of natural laws, it demonstrates a fundamental limitation of natural laws, namely, that they are incomplete. Think of it this way: There are lots and lots of things that happen in the world. For many of those things we can find causal antecedents that account for them in terms of natural laws. But why should everything that happens in the world submit to this sort of causal analysis? This is not to deny that all things happen for a reason. But the reason can be other than natural causes operating according to natural laws. The reason something happens can be that an intelligent agent willed it to happen. What's more, that intelligence need neither violate natural laws nor be reducible to them. (Certainly, Judeo-Christian theism holds that God is an intelligent agent who acts in absolute freedom and is unconditioned by any natural forces.)

Intelligent design regards intelligence as an irreducible feature of reality. Consequently, it regards any attempt to subsume intelligent agency under natural causes as fundamentally misguided and regards the natural laws that characterize natural causes as fundamentally incomplete. This is not to deny derived intentionality, in which artifacts, though functioning according to natural laws and operating by natural causes, nonetheless accomplish the aims of their designers and thus exhibit design. Yet whenever anything exhibits design in this way, the chain of natural causes leading up to it is incomplete and must presuppose the activity of a designing intelligence. Moreover, the scientific basis for such incompleteness is always information ex nihilo, where the information in question is specified complexity and cannot be reduced to the constitution and dynamics of the natural world.

The idea that nature is a closed system of natural causes and that natural causes provide a complete account of everything that occurs in nature is deeply entrenched in the West. In its current incarnation it is perhaps most directly traceable to Baruch Spinoza. Nevertheless, the idea that natural causes are complete has no more warrant than the idea that mathematics should be complete in the sense that every true mathematical claim should be deducible from a simple set of axioms. Kurt Gödel effectively demolished the latter misconception. The theory of intelligent design challenges the former. Moreover, it challenges that misconception by pointing to phenomena in nature that nature is in principle incapable of accounting for strictly in terms of natural causes, namely, phenomena that exhibit specified complexity.

Many theologians, going right back to the founder of liberal Christian theology (i.e., Friedrich Schleiermacher, who took his inspiration from Spinoza), find the incompleteness of nature theologically objectionable. Accordingly, for natural causes to lack the power to effect some aspect of nature would mean that God had not "fully gifted" the creation—this is Howard Van Till's way of putting it. Alternatively, a creator or designer who must act in addition to natural causes to produce certain effects has denied the creation benefits it might otherwise possess. Van Till portrays his God as supremely generous whereas the God of the design theorists he portrays as a miser. Van Till even refers to intelligent design as a "celebration of gifts withheld."

But there's a different way to see intelligent design theologically. Granted, if the universe is like a clockwork (as with the design arguments of the British natural theologians), then it would be inappropriate for God, who presumably is a consummate designer, to intervene periodically to adjust the clock. Instead of periodically giving the universe the gift of "clock-winding and clock-setting," God should simply have created a universe that never needed winding or setting. But what if instead the universe is like a musical instrument? Then it is entirely appropriate for God to interact with the universe by introducing design or, in this analogy, by skillfully playing a musical instrument. This analogy, far from being a quirky, too-whimsical newcomer to the design debate, actually predates the clock analogy. Consider the design arguments of the church fathers, like Gregory of Nazianzus, who likened the universe to a lute.

Change the metaphor from a clockwork to a musical instrument, and the charge that intelligent design requires withholding gifts from nature dissolves. So long as there are consummate pianists and composers, player pianos will always remain inferior to real pianos. The incompleteness of the real piano taken by itself is therefore irrelevant here. Musical instruments require a musician to complete them. Thus, if the universe is more like a musical instrument than a clock, it is appropriate for a designer to interact with it in ways that affect its physical state. On this view, for the designer to refuse to interact with the world is to withhold gifts.

20

NATURE'S RECEPTIVITY TO INFORMATION

What must nature be like for a designing intelligence to interact coherently with the world and generate the specified complexity we see in living things?

SUCH QUESTIONS HAVE BEEN PARTICULARLY TROUBLING to the scientific community since the time of René Descartes. Descartes was a substance dualist. As such, he divided reality into two primary substances, the physical and the spiritual (or nonphysical). The fundamental problem facing Descartes's dualism is how to get the physical and the spiritual to interact coherently. For Descartes, the physical world consisted of extended bodies that interacted via direct contact. Thus for a spiritual dimension to interact with the physical world could only mean that the spiritual caused the physical to move. In claiming that human beings consist of both spirit and matter, Descartes therefore had to argue for a point of contact between spirit and matter. He settled on the pineal gland because it was the one place in the brain where, as far as he could see, symmetry was broken and everything converged. (Most parts of the brain have right and left counterparts.)

Although Descartes's argument does not work, the problem it tries to solve is still with us and surfaces regularly in discussions about intelligent design. The physical world consists of physical stuff, and for a nonphysical designer to influence the arrangement of physical stuff seems to require that the designer intervene in, meddle with or in some way coerce this physical stuff. What is wrong with this picture of supernatural action by a designer? The problem is not a flat contradiction with the results of modern science. Take, for instance, the law of conservation of energy. Although the law is often stated in the form "energy can be neither created nor destroyed," in fact all we have empirical evidence for is the much weaker claim that "in an isolated system energy remains constant." Thus a supernatural action that moves particles or creates new ones is beyond the power of science to disprove because one can always argue that the system under consideration was not isolated.

There is no logical contradiction here. Nor is there necessarily a god-of-the-gaps problem here. It is certainly conceivable that a supernatural agent could act in the world by moving particles so that the resulting discontinuity in the chain of physical causality could never be removed by appealing to purely physical mechanisms (i.e., physical causes completely characterized by well-defined natural laws—these can be deterministic or indeterministic). The "gaps" in the god-of-the-gaps objection are meant to denote gaps of ignorance about underlying physical mechanisms. But there is no reason to think that all gaps must give way to ordinary physical explanations once we know enough about the underlying physical mechanisms. The mechanisms may simply not exist. Some gaps might constitute ontic discontinuities in the chain of physical causes and thus remain forever beyond the capacity of physical mechanisms.

In discussing the inadequacy of physical mechanisms to bring about design, we need to be clear that intelligent design is not wedded to the same positivism and mechanistic metaphysics that drives Darwinian naturalism. It's not that design theorists and Darwinian naturalists share the same conception of nature but then simply disagree about whether a supernatural agent sporadically intervenes in nature. In fact, intelligent design does not prejudge the nature of nature—that's for the evidence to decide. Intelligent design's tools for design detection, for instance, might fail to detect design. Even so, if intelligent design is so free of metaphysical prejudice, why does it continually emphasize mechanism? Why is it constantly looking to molecular machines and focusing on the mechanical aspects of life? If intelligent design treats living things as machines, then isn't it effectively committed to a mechanistic metaphysics, however much it might want to distance itself from that metaphysics otherwise?

Such questions confuse two senses of the term *mechanism*. Michael Polanyi noted the confusion back in the 1960s (see his article "Life Transcending Physics and Chemistry" in the August 1967 issue of *Chemical and Engineering News*):

> Up to this day one speaks of the mechanistic conception of life both to designate an explanation of life in terms of physics and chemistry [what I was calling *physical mechanisms*], and an explanation of living functions as machineries—though the latter excludes the former. The

term "mechanistic" is in fact so well established for referring to these
two mutually exclusive conceptions, that I am at a loss to find two
different words that will distinguish between them.

For Polanyi, mechanisms, conceived as causal processes operating in nature, could not account for the origin of mechanisms, conceived as "machines or machinelike features of organisms."

Hence in focusing on the machinelike features of organisms, intelligent
design is not advocating a mechanistic conception of life. To attribute such
a conception of life to intelligent design is to commit a fallacy of composition. Just because a house is made of bricks doesn't mean that the house
itself is a brick. Likewise just because certain biological structures can
properly be described as machines doesn't mean that an organism that includes those structures is a machine. Intelligent design focuses on the machinelike aspects of life because those aspects are scientifically tractable
and are precisely the ones that opponents of design purport to explain by
physical mechanisms. Intelligent design proponents, building on the work
of Polanyi, argue that physical mechanisms (like the Darwinian mechanism of natural selection and random variation) have no inherent capacity
to bring about the machinelike aspects of life.

But how, then, does an unembodied designer bring about the machinelike aspects of life and interact with nature generally except by overriding
the physical mechanisms by which nature operates and thus by directly
moving particles? Although an unembodied designer who moves particles is not logically incoherent, such a designer nonetheless remains problematic for science. The problem is that physical mechanisms are fully capable of moving particles. Thus for an unembodied designer also to move
particles can only seem like an arbitrary intrusion. The designer is merely
doing something that nature is already doing, and even if the designer is
doing it better, why did the designer not make nature better in the first
place so that it can move the particles better?

But what if the designer is in the business not of moving particles but of
imparting information? In that case nature moves its own particles, but an
intelligence nonetheless guides the arrangement that those particles assume. A designer in the business of moving particles accords with the following world picture: the world is a giant billiard table with balls in motion, and the designer arbitrarily alters the motion of those balls or even
creates new balls and then interposes them among the balls already
present. On the other hand, a designer in the business of imparting information accords with a very different world picture: in that case the world

becomes an information-processing system that is responsive to novel information. Now, the interesting thing about information is that it can lead to massive effects even though the energy needed to represent and impart the information can be infinitesimal. For instance, the energy requirements to store and transmit a launch code are minuscule, though getting the right code can make the difference between rocket boosters blasting a multi-ton shuttle into space or simply sitting there without making a peep. Frank Tipler and Freeman Dyson have even argued that arbitrarily small amounts of energy are capable of sustaining information processing and, in fact, of sustaining it indefinitely.

When a system is responsive to information, the dynamics of that system will vary sharply with the information imparted and will be largely immune to purely physical factors (e.g., mass, charge or kinetic energy). Consider the case of a steersman who guides a ship by controlling its rudder. The energy imparted to the rudder is minuscule compared to the energy inherent in the ship's motion, and yet the rudder guides its motion. It was this analogy that prompted Norbert Wiener to introduce the term *cybernetics*, which is derived etymologically from the Greek and means "steersman." It is no coincidence that in his book *Cybernetics*, Wiener writes about information as follows: "Information is information, not matter or energy. No materialism which does not admit this can survive at the present day."

How much energy is required to impart information? We have sensors that can detect quantum events and amplify them to the macroscopic level. What's more, the energy in quantum events is proportional to frequency or inversely proportional to wavelength. And since there is no upper limit to the wavelength of, for instance, electromagnetic radiation, there is no lower limit to the energy required to impart information. In the limit, a designer could therefore impart information into the universe without inputting any energy at all.

But limits are tricky. To be sure, an embodied designer could impart information by employing arbitrarily small amounts of energy. But an arbitrarily small amount of energy is still a positive amount of energy, and any designer employing positive amounts of energy to impart information is still in the business of moving particles. The question remains how an unembodied designer can influence the natural world without imparting any energy whatsoever. It is here that an indeterministic universe comes to the rescue. Although we can thank quantum mechanics for the widespread recognition that the universe is indeterministic, indeterminism has a long philosophical history, and appears in such diverse places as the atomism of Lucretius and the pragmatism of Charles Peirce and William James.

Our best physical evidence confirms that we live in an indeterministic world. (Even the multiworld interpretation of quantum mechanics does not recover determinism for this world but only for an ensemble of worlds.) What's more, given the imprecision inherent in all our measurements, there is no way ever to establish determinism with finality. Now, an indeterministic world will produce indeterministic events, and these could serve as a zero-energy conduit of intelligent design (i.e., for events exhibiting specified complexity that stand out against the backdrop of randomness). Since specified complexity is a reliable empirical marker of actual design, and since quantum mechanics powerfully suggests that all natural processes possess an indeterministic component, our best empirical evidence and scientific understanding of the world therefore confirms that we live in an indeterministic universe that is open to novel information. Consequently, that information could exhibit specified complexity (in biological systems) and therefore offer convincing evidence that an unembodied designer has imparted information. Notice the significance of this last step. An unembodied designer lends itself better than an embodied designer to design working through indeterminacy, for the embodied designer would be implicated in and a servant of the very indeterministic processes it sought to inform.

To see how an unembodied intelligence might impart information without imparting energy, consider a device that outputs zeroes and ones and for which our best science tells us that the bits are independent and identically distributed so that zeroes and ones each have probability 50 percent. (The device is therefore an idealized coin-tossing machine; note that quantum mechanics offers such a device in the form of photons shot at a polarizing filter whose angle of polarization is 45 degrees in relation to the polarization of the photons—half the photons will go through the filter, counting as a "one"; the others will not, counting as a "zero.") Now, what happens if we control for all possible physical interference with this device, and nevertheless the bit string that this device outputs yields an English text-file in ASCII code that delineates the cure for cancer (and is thus a clear instance of specified complexity)?

Given this setup, we have precluded that a designer imparted a positive amount of energy (however minuscule) to influence the output of the device. Nevertheless, there is no way to avoid the conclusion that a designer (presumably unembodied) influenced the output of the device despite imparting no energy to it. Note that there is no problem of counterfactual substitution here. It is not that the designer expended any energy and therefore did something physically discernible to the device in question. Any bit when viewed in isolation is the result of an irreducibly chance-driven process. And yet the arrangement and coordination of the bits in sequence cannot reasonably be attributed to chance and in fact points unmistakably to an intelligent designer.

It is here that critics of design typically throw up their hands in despair and charge that design theorists are merely evading the issue of how a designer introduces design into the world. Surely there must be some physical mechanism by which the information is imparted. Surely there are thermodynamic limitations governing the flow of information. Thermodynamic limitations do apply if we are dealing with embodied designers who need to output energy to transmit information. But unembodied designers who co-opt random processes and induce them to exhibit specified complexity are not required to expend any energy. For them the problem of expending energy to move material objects simply does not arise. Indeed, they are utterly free from the charge of counterfactual substitution, in which natural laws dictate that material objects would have to move one way but ended up moving another because an unembodied designer intervened. Indeterminism means that an unembodied designer can substantively affect the structure of the physical world by imparting information without imparting energy.

Nor does linking indeterminism to the action of an unembodied designer constitute a rejection of the principle of sufficient reason. This principle states that there's a sufficient reason or cause for the existence of any entity. Indeterminism is thought to deny this principle by suggesting acausality, where an event that is attributed to a random or indeterministic process is regarded as having no cause or at best an incomplete cause (i.e., whatever we are calling a cause does not provide a complete account of the event in question). If one views chance as fundamental and specified complexity as an anomaly of chance, then this view follows.

But one can also view design (i.e., the activity of intelligent agency) as fundamental and treat chance as a byproduct of design. To see how this might work, notice that agents can do things that follow well-defined probability distributions. For instance, even though I act as a designing agent in writing this book, the distribution of letter frequencies in it follows a well-defined probability distribution: the relative frequency of the letter *e* is approximately 13 percent, that of *t* is approximately 9 percent, and so on. The book itself is the result of design, but it also exhibits chance in the distribution of its letters. Even so, design here is more fundamental than chance: the design hypothesis identifying me as author of this book—an early twenty-first-century English-language author without any quirky need to overwork some particular letter or other in my prose—confers a certain probability distribution on the book's letter frequencies. Thus, in this instance, intelligent agency provides a sufficient reason for chance events.

One can take this line of reasoning further and argue that chance and randomness do not even make sense apart from design. (I argued this in a paper titled "Randomness by Design" that appeared in *Nous* in 1991.) Consider that for any chance process like tossing a coin, if the coin is tossed indefinitely, any finite sequence will not only appear once but also appear infinitely often. (This follows from the Strong Law of Large Numbers.) Now consider further that we can have experience only of finite sequences of coin tosses. Suppose, therefore, that I come in at some arbitrary point in the tossing of a coin that is being tossed indefinitely. On what basis can I have confidence that the finite sequence I witness will in some way be "representative of chance"?

For instance, on what basis should I expect to see approximately the same number of heads and tails? Since the coin is being tossed indefinitely, even if I witness a million coin tosses, I could encounter a run of a million heads in a row. What keeps me from coming in at the beginning of such a run and witnessing a million heads in a row? To be sure, one can argue that it is highly unlikely that I'll come in at the precise point where I witness a million heads in a row. But that merely restates the problem. It does not resolve it. To say that chance will behave representatively of chance because the chances are that it will do so is no answer because we can well imagine a world in which irreducible chance processes with inherent propensities for producing certain probabilistic behavior never exhibit behavior representative of those propensities. (Imagine a world exactly like ours, but in which fair coins—i.e., rigid homogeneous disks with distinguishable sides—are always observed to land heads 90 percent of the time.)

But suppose instead that chance and randomness are byproducts of design. Then the problem of chance inducing events that are unrepresentative of chance (e.g., a million heads in a row) can be effectively circumvented since design has a way of stabilizing and thereby justifying chance. To come back to the letter frequencies of English texts, although those frequencies follow a chance distribution, they are perfectly stable as a result of orthographic, syntactic, grammatical and semantic constraints on English. For those letter frequencies to diverge sharply from the norm would therefore itself constitute an instance of design. In 1939 Ernest Vincent Wright published a novel of over fifty thousand words titled *Gadsby* that contained no occurrence of the letter *e*. Clearly, the absence of the letter *e* was designed.

There is no logical inconsistency and no evading the hard problems of science in treating the world as a medium receptive to information from an unembodied intelligence. In requiring a mechanistic account of how an unembodied intelligence imparts information and thereby introduces design into the world, the design critic exhibits a failure of imagination. Such a critic is like a physicist trained only in Newtonian mechanics who is desperately looking for a classical account of how a single particle like an electron can go through two slits simultaneously to produce a diffraction pattern on a screen. (This is the famous double-slit experiment; technically, the diffraction pattern is a statistical distribution that results from multiple electrons being shot individually at the screen.)

On a classical Newtonian view of physics, only a classical account in terms of sharply localized and individuated particles makes sense. And yet nature seems unwilling to oblige any such account of the double-slit experiment. Richard Feynman was right when he remarked that no one understands quantum mechanics. The "mechanics" in "quantum mechanics" is nothing like the "mechanics" in "Newtonian mechanics." There are no analogies that carry over from the dynamics of macroscopic objects to the quantum level. In place of understanding we must content ourselves with knowledge. We do not *understand* how quantum mechanics works, but we *know* that it works. So too, we may not *understand* how an unembodied designer imparts specified complexity into the world, but we can *know* that such a designer imparts specified complexity into the world.

Even so, there is a useful way to think about how an unembodied designer imparts information into the world. In the liturgies of most Christian churches, the faithful pray that God keep them from sinning in "thought, word and deed." Thoughts left to themselves are inert and never accomplish anything outside the mind of the individual who thinks them. (Design merely in the mind of God or the believer is scientifically useless.) Deeds, on the other hand, are coercive, forcing physical stuff to move now this way and now that way. (It is no accident that the concept of force plays such a crucial role in the rise of modern science.) But between thoughts and deeds are words. Words mediate between thoughts and deeds. Words give expression to thoughts and bring the speaker into communion with the hearer. On the other hand, words by themselves are never coercive; without deeds to back them up, words lose their power to threaten. Nonetheless, words have the power to engender deeds by finding a receptive medium that can then act on them. The world is such a receptive medium.

21

THE LAW OF CONSERVATION OF INFORMATION

What does it mean to say that specified complexity or complex specified information is conserved?

IN *THE LIMITS OF SCIENCE*, PETER MEDAWAR defines a conservation law for deterministic processes that he calls "the Law of Conservation of Information." According to this law, when a deterministic process operates (be it a deterministic natural process or a deterministic computer algorithm), the amount of information that the process outputs can never exceed the amount of information with which it started. We might state it this way: *Deterministic processes cannot generate information.*

To take a simple example, suppose you want a copy of an out-of-print book that's in your school library. You therefore check out the book and photocopy it. The photocopy machine acts as a deterministic process (or at least approximately so) that outputs a copy of the book. You then keep the copy and return the original to the library. In so doing, you have no more information than you started with—the copy contains no more information than the original. Nor for that matter do both copies together contain more information than any one copy individually. The mathematics justifying these claims is straightforward.

Now Medawar's version of the Law of Conservation of Information was deterministic and thus didn't address either chance or the joint action of chance and necessity. Neither could his version be straightforwardly generalized to the nondeterministic case. The problem is that chance can generate plenty of novel information. (Just get out your coin and start tossing it. After a few dozen tosses, the sequence of heads and tails that you witness will be unique in the history of coin tossing and constitute novel information.) But what chance—whether pure chance or chance combined with necessity—can't generate is the particular type of information that is the focus of intelligent design, namely, specified complexity. We may therefore formulate the following nondeterministic version of the Law of Conservation of Information: *Neither chance, nor necessity nor their combination is able to generate specified complexity or, equivalently, complex specified information.* And since chance, necessity and the combination of these constitute what is typically meant by natural causes, the upshot of this law is that natural causes are incapable of generating specified complexity.

So stated, the Law of Conservation of Information at once generalizes and restricts Medawar's deterministic version of it. On the one hand, Medawar's version applies only to necessitarian processes whereas the nondeterministic version applies both to chance and necessitarian processes. On the other hand, Medawar's version applies to information generally (specified or unspecified, complex or noncomplex) whereas the nondeterministic form applies only to a particular kind of information— specified complexity or, equivalently, complex specified information.

The Law of Conservation of Information tells us nothing fundamentally new about specified complexity. In a sense, it merely restates what the design inference has told us all along, namely, that you can't get specified complexity through chance, necessity or their combination. Even so, the law is useful for keeping track of where specified complexity enters and exits a natural process. The key word in the definition of the Law of Conservation of Information is *generate*. Natural processes are quite adept at shuffling preexisting specified complexity. But what they can't do is generate it from scratch. The problem is that if this shuffling of preexisting specified complexity is sufficiently intricate, it can give the appearance of generating specified complexity from scratch. The Law of Conservation of Information is essentially a bookkeeping device: it keeps us honest about the sources of specified complexity. In particular, when specified complexity seems to be generated by purely natural causes, the law helps us to identify how in fact it was front-loaded, smuggled in or hidden from view.

How is the Law of Conservation of Information a *conservation* law? Ordinarily, when something is conserved, some quantity characterizing that thing remains unchanged. This is certainly the case with conservation of energy, in which the energy of an isolated system remains perfectly constant. Information, however, is rarely conserved in the sense of remaining constant. In our ordinary experience, information can increase under the operation of natural causes—for instance, random coin tossing generates information (though not specified complexity). Alternatively, information can decrease under the operation of natural causes—for instance, erosion can destroy the information in a statue (in that case losing specified complexity). The Law of Conservation of Information tells us that when specified complexity is given over to natural causes, it either remains unchanged (in which case information is strictly conserved) or disintegrates (in which case information diminishes).

For instance, the best thing that can happen to a book on a library shelf is that it remains as it was when originally published and thus preserves the specified complexity inherent in its text. Over time, however, what usually happens is that a book gets old, its pages fall apart, and the information on the pages disintegrates. The Law of Conservation of Information is therefore more like a thermodynamic law governing entropy than a conservation law governing energy, with the focus on degradation rather than conservation. Yet because reference to a conservation law already has some currency (due to Peter Medawar) and also because the law says that we can do no better than break even (implying conservation in the ideal case), it seems best to refer to this law as a conservation law. The crucial point of the Law of Conservation of Information is that natural causes (conceived as chance, necessity and their combination) can at best preserve specified complexity, or they may degrade it, but they cannot generate it.

It is important to be clear about just what is being denied when the Law of Conservation of Information claims that natural causes cannot generate specified complexity. To deny that they can generate specified complexity is not the same as to deny that they can produce events that exhibit specified complexity. Natural causes are entirely suitable as conduits for specified complexity. (Think of a photocopy machine that takes preexisting specified complexity and, by an automatic process, produces a copy that likewise exhibits specified complexity.) Natural causes may thus be said to "produce" specified complexity or complex specified information.

Even so, natural causes never produce things *de novo* or ex nihilo. Whenever they produce things, they do so by reworking other things. Thus, to the question, "How did natural causes produce X?" it is never enough to assert that natural causes simply did it. Rather, one must point to some antecedent Y that is causally sufficient to account for X. This is the case regardless of whether the natural cause that produced X operated deterministically or nondeterministically. Gravity, operating deterministically, is sufficient to account for the falling of a metal ball near the earth's surface. Radioactivity, operating nondeterministically, is sufficient to account for the decay of uranium into lead and helium.

The sufficiency of natural causes to produce an effect is crucial to naturalistic explanations since without it the door lies open to intelligent agency. Naturalistic explanations by definition exclude appeals to intelligent agency. Thus, to say that a natural cause produced X and then to point to some antecedent Y and say that, under the operation of natural causes, Y only contributed to the production of X is not enough for a bona fide naturalistic explanation. Such an explanation leaves the door open to intelligent agency acting in tandem with natural causes. A mixture of intelligent and natural causes is clearly not being denied when the Law of Conservation of Information states that specified complexity cannot be generated by natural causes. The Law of Conservation of Information is not saying that natural causes in tandem with intelligent causes cannot generate specified complexity but rather that natural causes apart from intelligent causes cannot generate specified complexity. Thus, to attribute X to natural causes is a call for an explanation in terms of some antecedent circumstance Y upon which natural causes—and only natural causes—operate and suffice to produce X. The Law of Conservation of Information says that if X exhibits specified complexity, then so does Y. It follows that natural causes do not, and indeed cannot, generate specified complexity but merely shuffle it around.

The Law of Conservation of Information has profound implications for science. Among its immediate corollaries are the following:

1. The specified complexity in a closed system of natural causes remains constant or decreases.

2. Specified complexity cannot be generated spontaneously, originate endogenously or organize itself (as these terms are used in origins-of-life research).

3. The specified complexity in a closed system of natural causes either has been in the system forever or was at some point added exogenously (implying that the system, though now closed, was not always closed).

4. In particular, any closed system of natural causes that is also of finite duration received whatever specified complexity it contains before it became a closed system.

The first corollary can be understood in terms of data storage and retrieval. Data often constitute a form of specified complexity. Ideally, data would stay unaltered over time. Nonetheless, entropy being the corrupting force that it is, data tend to degrade and need constantly to be restored.

Over time magnetic tapes deteriorate, pages yellow, print fades and books fall apart. Information may be eternal, but the physical media that house information are subject to natural causes and are thoroughly ephemeral. The first corollary acknowledges this fact.

Given an instance of specified complexity, corollaries (2) and (3) allow but two possibilities to explain it: either the specified complexity was always present or it was inserted. Proponents of intelligent design differ about which of these two possibilities obtains. This debate is not new. The German teleomechanists and the British natural theologians engaged in a similar debate, with the Germans arguing that teleology was intrinsic to the world (and therefore ever-present) and the British arguing that it was extrinsic (and therefore inserted at various times). However such debates get resolved, specified complexity is an empirically detectable entity that is not reducible to natural causes (which, as usual, are understood nonteleologically as chance, necessity and their combination).

Corollary (4) implies that scientific explanation is not identical with reductive explanation. Richard Dawkins, Daniel Dennett and many other scientists and philosophers are convinced that proper scientific explanations must be reductive, explaining the complex in terms of the simple. The Law of Conservation of Information, however, shows that specified complexity cannot be explained reductively. To explain an instance of specified complexity requires either appealing to an intelligent agent that originated it or locating an antecedent instance of specified complexity that contains at least as much specified complexity as we are trying to explain. A pencil-making machine is more complicated than the pencils it makes. A clock factory is more complicated than the clocks it produces. What's more, all causal chains from pencil to pencil-making machine or from clock to clock factory ultimately trace back to intelligence. Intelligent causes generate specified complexity; natural causes merely transmit preexisting specified complexity (and usually do so imperfectly).

Thus, to explain specified complexity in terms of natural causes is to fill one hole by digging another. With specified complexity the information problem never goes away, short of locating the intelligence that originated it. We have known this since elementary school. The telephone game—where one person originates a message and then whispers it to the next person, who whispers it to the next person, and so on—illustrates how information originates and degrades. The players of this game are links in a chain. With each transmission of information from one link to the next, there is the potential for losing information. Ideally, each person would repeat exactly the information given by the preceding person in the chain and thus preserve the information given at the start. In general, however, that does not happen. In fact, the fun of the telephone game is to see how information degrades as it passes from the first to the last person in the chain.

The telephone game has more serious analogues. Consider the textual transmission of ancient manuscripts. A textual critic's task is to recover as much of the original text of an ancient manuscript as possible. Almost always the original text is unavailable. Instead, the textual critic confronts multiple variant texts, each with a long genealogy tracing back to the original text. Fifty generations of copies may separate a given manuscript from the original text. The original text is copied in the first generation, then that copy is itself copied, then that second copy is in turn copied, and so on fifty times before we get to the manuscript in the critic's possession. We assume that most of the copyists were trying to preserve the text faithfully. Even so, they were bound to introduce errors now and then. Worse yet are the wayward copyists who use copying as a pretext for inserting their own pet ideas into a text. The textual critic must therefore identify errors introduced by careless copying as well as errors stemming from a copyist's personal agenda. This can be enormously difficult. Even so, there is always a fixed reference point: because all the variant manuscripts ultimately spring from the same original text, the original text constitutes the initial specified complexity on which the transmission of texts depends.

In both the telephone game and the transmission of texts, an intelligent cause rather than a natural cause not only generates the initial specified complexity but then also transmits it. Is the role of intelligence in transmitting specified complexity a problem here, given that the Law of Conservation of Information applies, strictly speaking, only to the transmission of specified complexity by natural causes? Although this law is concerned solely with placing limits on natural and not intelligent causes, it still applies. Intelligent causes can mimic natural causes, and that is what they are doing here. In both the telephone game and in the transmission of ancient texts, the persons transmitting information are supposed to repeat what they have been given. Repetition is an automatic process for which natural causes are ideally suited and for which intelligent causes are not required.

For our purposes, the most interesting application of the Law of Conservation of Information is the reproduction of organisms. In reproduction, organisms transmit their specified complexity to the next generation. For most evolutionary biologists, however, there's more to the story.

Most evolutionists hold that the Darwinian mechanism of natural selection and random variation introduces novel specified complexity into an organism, supplementing the specified complexity that parents transmit to their offspring with specified complexity from the environment. In contrast, the Law of Conservation of Information makes clear that without prior input of specified complexity into the environment, no natural mechanism, not even the Darwinian mechanism, will be able to produce specified complexity.

PART 4

ISSUES ARISING
FROM NATURALISM

22

VARIETIES OF NATURALISM

Is naturalism in any guise compatible with intelligent design?

NATURALISM COMES IN FOUR MAIN GUISES: *antiteleological, methodological, antisupernaturalist* and *pragmatic*. Of these only the last two are compatible with intelligent design, and only the last is compatible with Christian theism. Naturalism, as the word suggests, places the spotlight on nature. Naturalism in all its guises is concerned with how we should properly understand nature.

Antiteleological naturalism is the predominant form of naturalism—it's what's usually meant by the term *naturalism*. Antiteleological naturalism takes nature to be all there is and views nature, at the nuts-and-bolts level, as operating purely by blind natural causes. These are causes characterized by chance and necessity and ruled by unbroken natural laws. Antiteleological naturalism leaves no room for any fundamental teleological principles operating in nature. It therefore rules out intelligent design not merely for being scientifically problematic but also for having no conceivable purchase on reality. If antiteleological naturalism is true, then there can be no fundamental design or teleology inherent in nature. Any such design or teleology must evolve as a result of more basic laws controlled ultimately by chance and necessity. Jacques Monod, in *Chance and Necessity*, embraced antiteleological naturalism and made it a fundamental principle for science: "The cornerstone of the scientific method is the postulate that nature is objective. In other words, the *systematic* denial that 'true' knowledge can be got at by interpreting phenomena in terms of final causes—that is to say, of 'purpose.' "

Methodological naturalism accepts Monod's statement as a regulative principle for science but, contra Monod, is willing to grant that there may be more to reality than chance and necessity. Methodological naturalism doesn't care what you believe deep down. Yet for the sake of science, methodological naturalism insists that scientists pretend as though antite-leological naturalism is true. Nancey Murphy has called this view *method-ological atheism*. The idea here is that science is a method for investigating nature and that to understand nature scientists must only invoke "natural processes." In this context the term "natural processes" means processes operating entirely according to unbroken natural laws and characterized by chance and necessity.

As a regulative principle for science, methodological naturalism is a rule for keeping science on the straight and narrow by preventing it from veering into the supernatural or occult. One of the consequences of meth-odological naturalism is to exclude intelligent design from science. Teleol-ogy and design, though perhaps real in some metaphysical sense, are, ac-cording to methodological naturalism, not proper subjects for inquiry in the natural sciences. To speak of the empirical detectability of intelligent design is therefore an oxymoron from the vantage of methodological nat-uralism. Something is empirically detectable and thus open to scientific in-quiry only if it results from natural processes that by definition render de-sign and teleology empirically undetectable.

For the working scientist, antiteleological and methodological natural-ism are effectively equivalent. Although the antiteleological naturalist is an atheist (like Daniel Dennett or Richard Dawkins) and the methodolog-ical naturalist can be a theist (like many Christians in the American Scien-tific Affiliation), their science is indistinguishable. The antiteleological nat-uralist may think that Darwin hit the nail on the head in explaining the evolution of life by natural selection. The methodological naturalist may likewise think that Darwin hit the nail on the head in explaining the evo-lution of life by natural selection but then, if also a theist, holds that Dar-winian evolution was God's method of creating life. But understand, for the methodological naturalist who is also a theist, neither God nor design nor teleology plays any substantive role in the natural sciences. The meth-odological naturalist takes evolution as God's method of creating life but rules out of court the possibility that God might have left any empirical fingerprints in using such a method of creation.

Although methodological naturalism is a regulative principle that pur-ports to keep science on the straight and narrow by limiting science to nat-ural causes, in fact it is a straitjacket that actively impedes the progress of science. If methodological naturalism were merely a working hypothesis, maintained because it supposedly has served science well in the past, that would be one thing. As a working hypothesis, it would be optional, and scientists who found the hypothesis no longer helpful would be free to discard it. But methodological naturalism isn't saying that we have yet to encounter empirical evidence of design in nature but we should stay open to it in case it comes along. Rather, methodological naturalism insists that one is most logical, most scientific, if one pretends such an empirical possibility is logically impossible. Instead of holding methodological naturalism as a working hypothesis, methodological naturalists hold it as a dogma.

The fact is, there are plenty of ways nature might offer empirical confirmation for intelligent design. If, for instance, every bacterium produced a plasmid (a circular piece of DNA) that encoded as an ASCII file a unique limerick (i.e., a different limerick for each individual bacterium, and thus trillions upon trillions of limericks), then there would be no question that such DNA (the "limerick plasmid") was designed. The identity of that designer might be a matter of debate but the fact of design would not. To be sure, finding such design seems wildly implausible, but methodological naturalism holds that if science did encounter evidence for such design, it would have to ignore it or else dismiss it as the work of an evolved designer whose evolution did not involve any evident design. Only designers reducible to nondesigners are acceptable within methodological naturalism. Granted, the limerick plasmid exhibits flamboyant design whereas the actual designs in biology appear far more subtle. But does that make the actual designs any less real? Or, for that matter, does it preclude biological design from being empirically detectable? Of course not.

Methodological naturalism encourages science to continue business as usual by restricting itself solely to blind natural causes and the unbroken natural laws that describe them. But why should science continue business as usual? How do we know that empirical inquiry into the natural world can only uncover the effects of chance and necessity but not design? How do we know that blind natural causes are the best that science has available in studying nature? Clearly, the only way to answer these questions is to go to nature with an open mind and see whether nature exhibits things that require a designing intelligence. Intelligent design leaves that possibility open without deciding it in advance. Methodological naturalism, by contrast, excludes the possibility of design apart from any consideration of evidence.

Methodological naturalism is a rule, not a metaphysical position. In contrast, both antiteleological and antisupernaturalist naturalism are metaphysical positions. Antiteleological naturalism is at the heart of most varieties of naturalism, including philosophical naturalism, metaphysical naturalism, epistemological naturalism, reductive naturalism, scientific naturalism, scientific materialism, materialism and physicalism. (Under this last designation fall both reductive and nonreductive physicalism. The emergence of nonreductive physicalism is yet another attempt to explain away any fundamental or irreducible teleology.) If one speaks of naturalism simpliciter (i.e., naturalism without a qualifying adjective in front), one typically means antiteleological naturalism. Naturalists, then, are those who hold to this form of naturalism.

The third form of naturalism we need to consider is antisupernaturalist naturalism. It arises from recent efforts to reform naturalism and open it to a real religious dimension (one that is substantial and not just a Freudian projection or wish fulfillment). This reformed naturalism, which may also be called *religious naturalism*, is as much a metaphysical position as antiteleological naturalism. Perhaps the chief spokesperson for religious naturalism is philosopher of religion and process theologian David Ray Griffin. He refers to this position as naturalism$_{ns}$ (the subscript *ns* here refers to "nonsupernaturalist") and contrasts it with what he calls naturalism$_{sam}$ (the subscript *sam* here refers to "sensationist-atheist-materialist"), which is essentially what I'm calling antiteleological naturalism. Howard Van Till, also a proponent of religious naturalism, seems to prefer the term *theistic naturalism*. (Note well what is the adjective and what is the noun here.)

The animating impulse behind antisupernaturalist or religious naturalism is twofold: first, to open reality to a richer set of possibilities than is available through antiteleological naturalism (and thereby to ground genuine religious experience, which typically presupposes some ultimate purposiveness underlying reality); and, second, to keep reality firmly anchored to the principles that govern nature. Griffin's writings are instructive in this regard. He sees antiteleological naturalism as unable to account for life after death, paranormal phenomena and the emergence of biological complexity. The world, as he sees it, operates according to a richer set of principles than chance and necessity; and these principles, according to him, make possible life after death, paranormal phenomena and the emergence of biological complexity. Nevertheless, these principles, though richer than the laws and causal principles of antiteleological naturalism, are inviolable. God (a process God who evolves with the world, in Griffin's case) is himself bound by these principles. Thus Griffin will stress over and over again that the one thing religious naturalism proscribes is supernatural intervention (hence his continual emphasis on "nonsupernaturalist naturalism"). Supernatural intervention would require God to violate the causal principles that govern reality.

From this characterization of religious or antisupernaturalist natural-ism we can see why this view is at once compatible with intelligent design but incompatible with Christian theism. As for being compatible with in-telligent design, since the world of the religious naturalist operates accord-ing to principles that are richer than chance and necessity, it could well in-clude teleological principles. The effects of these principles might well be empirically distinguishable from the effects of chance and necessity. One such principle might be that nature generates specified complexity under certain circumstances or is programmed to express specified complexity at various points in its history.

These richer principles, however, do not allow supernatural intervention. For Griffin, for instance, there is no such thing as predictive prophecy. God evolves with the world and is rooted in time. Thus God cannot foreknow the future. At best, God can offer educated guesses about the future. Also, God, by evolving in time, is at every point in time required to work with the world as it is and thus in accord with the principles that govern the world. This allows God gently to beckon the world and gradually try to steer it in new directions. It does not, however, allow God to introduce radical discon-tinuities, like raising a man from the dead or creating species from scratch. Evolution, not revolution, is the watchword of process theology.

Process theology, and the religious naturalism that inspires it, views the world and the principles that undergird it as primary. God then comes sec-ond, being conditioned by those principles (much like the Greek gods in-habiting Mount Olympus, who were conditioned by what the Fates de-creed). Accordingly, God is a directing influence that guides the world but is constrained by these more basic principles. Yet with no veto power over these more basic principles, the process God must take what's given. In-deed, the process God always bows to the freedom of creation. By contrast, within classical theism, creation always bows to divine freedom.

The process view of God should not be surprising given the prove-nance of process theology. Alfred North Whitehead, the father of process theology and the inspiration for Griffin's religious naturalism, was a big fan of both Plato and a progressive form of evolution. Process theology is essentially a marriage of these views. For Plato, what's primary is not a personal creator but the forms or ideas, which are dimly reflected in na-ture. Thus, when the Platonic Demiurge creates (see Plato's *Timaeus*), he is constrained by both the forms and the preexisting materials of the world. To this picture Whitehead adds that the Demiurge's creation must take place as a gradual unfolding of an evolutionary process that is progressing to ever-higher degrees of novelty, beauty and enrichment.

This picture of reality has a lot going for it, as witnessed by the increasing popularity of process theology throughout the theological world. It certainly provides a richer set of possibilities than the impoverished ontology of antiteleological naturalism. Also, it promises to resolve certain longstanding theological conundrums. By making God subject to principles that govern the world, the problem of theodicy has a straightforward resolution: God is benevolent but not omnipotent. He means well, but a lot of evil is simply beyond his ability to prevent. Religious naturalism also promises to do away with the legacy of Cartesian dualism: by eliminating the supernatural it allows for a nondualistic interaction between the mental and physical conceived as upward and downward causation within a single causal nexus. (Within Cartesian dualism the mental and the physical were two completely different kinds of causal powers operating in two completely different kinds of causal nexuses—the material and the spiritual.)

For the Christian, there's just one problem with process theology and the religious naturalism that undergirds it: its doctrines of God and creation are totally unacceptable. Christian theology is not process theology. Christian theology, properly so-called, regards the doctrine of *creatio ex nihilo*, or *creation from no preexisting stuff*, as nonnegotiable. *Creatio ex nihilo* presupposes two things: that God is a personal being and not a principle, and that the world exists by a personal act, namely, an effected word spoken by God. The early theologians of the Christian church (like Athanasius and the Cappadocian fathers) were all too aware of Plato's cosmology. The problem, in their view, with Plato's cosmology was just that: namely, Plato's world was a cosmos, an ordered arrangement governed by principles that even God would have to obey. The Christian God, by contrast, was absolutely free, and the world, as an absolutely free act by this absolutely free God, was not a cosmos (at least not in the first instance) but a creation.

There is a logic here that's inescapable and that leads in either of two completely opposite directions. Either God is free or God is bound. Unless God is absolutely free and the world is an absolutely free act of creation, then there are principles that constrain God in creation. (The issue of God being bound by his nature is not a problem here so long as God's nature is not set over and against God. Thomas Aquinas, for instance, identified God's essence or nature with God's existence.) Any such principles of cosmic constraint, however, are logically prior to God. But in that case the ultimate reality is not God but those principles. Take your pick—either the ultimate reality is the one personal God or something else is the ultimate reality, like Plato's forms or Whitehead's process. The one choice leads to theism, the other to naturalism.

It's no accident that Griffin embraces naturalism. Naturalism for him bespeaks the principles that govern the world (and above all, the evolutionary process). That's what's ultimate for him. God within Griffin's process view is but one player in a world of other players; God is not the ultimate reality. (We see this repeatedly in contemporary theological discussions where the ultimacy of God gives way to some other ultimacy.) For Christian theism, by contrast, God has always been the ultimate reality. Yet as soon as God is in some way limited, what limits God becomes the ultimate reality.

The only way to maintain that God is the ultimate reality, therefore, is either to identify God with nature and the principles that govern nature (as in pantheism and the philosophy of Baruch Spinoza) or to allow for God to be an absolutely free personal being (as in orthodox Christian theism). Panentheism, the view that the world is intrinsically in God, is not an option here because it makes God depend on some deeper reality (like the principles governing the world). The choice, then, is to make either nature ultimate or divine personhood ultimate. Christian theism opts for the latter, making the personal, and not impersonal nature, God. Trinitarian theology locates the personhood of God within the interpersonal triune life. God the Father creates through the Son in the power of the Holy Spirit. By a free act that mirrors the intratriune personal relations, God creates a world of finite creatures (which include physical as well as spiritual beings, humans as well as angels).

I remarked earlier that process theology and the religious naturalism undergirding it resolves the theodicy problem by making God good but not sufficiently powerful to prevent evil. (The principles that govern nature in this case overrule God's good intentions.) I also pointed to some other advantages of religious naturalism over antiteleological naturalism, like a richer ontology of nature that could include real teleology. That said, I don't want to leave the impression that it's reasonable to say, "Wow, process theology is the most powerful explanatory tool in town. Bible thumpers just don't like it because it doesn't jibe with the Good Book."

Process theology's faulty doctrine of creation has some deeply unsatisfying theological implications. For instance, process theology leaves us with an existentially disturbing explanation for the apparent ontological difference between good and evil. (Within process theology, evil is simply a cost of nature's freedom.) Also, by presenting us with a God who means well but may not have the power to pull off his good intentions, process theology leaves us with no assurances for the future (except perhaps that God is trying his best and feels our pain). Process theologian Charles Hartshorne wrote a book titled *Omnipotence and Other Theological Mistakes*. To be sure, divine omnipotence raises difficult theological questions (e.g., why doesn't God prevent evil if he can), but divine impotence raises difficult theological questions as well (e.g., how can we have confidence that good will ultimately triumph if God simply may not have what it takes to bring it about).

Despite these important differences, antisupernaturalist naturalism, like Christian theism, leaves nature open to real teleology and so opens the door to intelligent design as a scientific enterprise. In contrast, antiteleological and methodological naturalism completely rule out intelligent design as a scientific project: intelligent design is a nonstarter if one adopts either of these views. Nevertheless, because antisupernaturalist naturalism is so much friendlier to intelligent design, one runs the risk of overlooking just how metaphysically uneasy is the fit between intelligent design and antisupernaturalist naturalism. The problem is that the whole concept of design implies giving something a capacity it did not possess before. As Aristotle put it, the art of ship building is not in the wood; it takes a designer to arrange the wood to make the ship. Design arranges preexisting materials and thereby confers on them something they did not previously possess.

But within antisupernaturalist naturalism, there is no bestowing of any gift on nature. Naturalism, whether antisupernaturalist or antiteleological, views nature as the ultimate reality and one that is complete in itself. Thus any designer, even the God of process theology, depends on some underlying principle or process of nature that inherently is not a designer. For this reason the language of design tends not to be very appealing to antisupernaturalist or religious naturalists. The language of purpose, teleology and even intelligence is okay. But just as Plato's Demiurge never played a terribly significant role in Plato's philosophy, so design can be expected to play a very minor role in religious naturalism and process theology (regardless of what scientific successes intelligent design might attain). By contrast, the God of Christianity is a designer. To be sure, Christianity's God is not merely a designer. But he is at least a designer.

There's one final form of naturalism that manages more than an uneasy relationship with Christian theism but rather is perfectly compatible with it—namely, pragmatic naturalism. Pragmatic naturalism wants simply to understand nature and doesn't care what entities are invoked to facilitate that understanding, so long as they prove conceptually fruitful. The philosopher Willard Quine was a pragmatic naturalist (as was Ludwig Wittgenstein). Accordingly, Quine was able to entertain the following possibility: "If I saw indirect explanatory benefit in positing sensibilia, possibilia, spirits, a Creator, I would joyfully accord them scientific status too, on a par with such avowedly scientific posits as quarks and black holes" ("Naturalism; or, Living within One's Means," *Dialectica* 1995, vol. 49). Quine's pragmatic naturalism clearly places no restraint on intelligent design or, for that matter, on Christian theism.

23

INTERVENTIONISM

*Is intelligent design an interventionist
theory in which design events punctuate an
otherwise fully natural causal history?*

INTELLIGENT DESIGN DOES NOT REQUIRE ORGANISMS to emerge suddenly
or to be specially created from scratch by the intervention of a designing
intelligence. To be sure, intelligent design is compatible with the creationist idea of organisms being suddenly created from scratch. But it is
also perfectly compatible with the evolutionist idea of new organisms
arising from old by a gradual accrual of change. What separates intelligent design from naturalistic evolution is not whether organisms
evolved or the extent to which they evolved but what was responsible for
their evolution.

Naturalistic evolution holds that material mechanisms alone are responsible for evolution (the chief of these being the Darwinian mechanism
of random variation and natural selection). Intelligent design, by contrast,
holds that material mechanisms are capable of only limited evolutionary
change and that any substantial evolutionary change would require input
from a designing intelligence. Moreover, intelligent design maintains that
the input of intelligence into biological systems is empirically detectable:
that is, it is detectable by observation through the methods of science. For
intelligent design, the crucial question therefore is not whether organisms
emerged through an evolutionary process or suddenly from scratch but
whether a designing intelligence made a discernible difference—regardless of how organisms emerged.

For a designing intelligence to make a discernible difference in the
emergence of some organism, however, seems to require that an intelligence intervened at specific times and places to bring about that organism and, thus again, seems to require some form of special creation. This
in turn raises the question: how often and at what places did a designing

intelligence intervene in the course of natural history to produce those biological structures that are beyond the power of material mechanisms? One of the criticisms of intelligent design is that it draws an unreasonable distinction between material mechanisms and designing intelligences, claiming that material mechanisms are fine most of the time but then on rare (or perhaps not so rare) occasions a designing intelligence is required to get over some hump that material mechanisms can't quite manage.

This criticism is misconceived. The proper question is not how often or at what places a designing intelligence intervenes but rather at what points do signs of intelligence first become evident. To understand the difference, imagine a computer program that outputs alphanumeric characters on a computer screen. The program runs for a long time and throughout that time outputs what look like random characters. Then, abruptly the output changes and the program outputs sublime poetry. Now, at what point did a designing intelligence intervene in the output of the program? Clearly, this question misses the mark because the program is deterministic and simply outputs whatever the program dictates.

There was no intervention at all that changed the output of the program from random gibberish to sublime poetry. And yet, the point at which the program starts to output sublime poetry is the point at which we realize that the output is designed and not random. Moreover, it is at that point that we realize that the program itself is designed. But when, where and how was design introduced into the program? Although these are interesting questions, they are ultimately irrelevant to the more fundamental question: was there design in the program and its output in the first place? Similarly, in biology there will be clear times and locations where we can say that design first became evident. But the precise activity of a designing intelligence at those points will require further investigation and may not be answerable. As the computer analogy just given indicates, the place and time at which design first becomes evident need have no connection with the place and time at which design was actually introduced.

Intelligent design is not a theory about the frequency or locality or modality by which a designing intelligence intervenes in the material world. It is not an interventionist theory at all. Indeed, intelligent design is perfectly compatible with all the design in the world coming to expression by the ordinary means of secondary causes over the course of natural history, much as a computer program's output comes to expression simply by running the program (and thus without monkeying with the program's operation). In fact, one way to think of the secondary causes responsible for biological evolution is as intelligently designed programs whose computational environment is the universe and whose operating system is the laws of physics and chemistry. This actually is an old idea, and one that Charles Babbage, the inventor of the digital computer, explored in the 1830s in his *Ninth Bridgewater Treatise* (thus predating Darwin's *Origin of Species* by twenty years).

Of course, there are other ways to think about secondary causes that leave room for genuine teleology in nature. Programming is one option, but it implies a highly mechanical or algorithmic view of secondary causation. Augustine, by contrast, thought of design in the world as coming to expression through seeds planted by God at creation. Here we have an organismic rather than algorithmic view of secondary causation. Physical necessity can also be the carrier of teleology through laws of form that channel evolution along certain preset paths. Orthogenesis from the late nineteenth and early twentieth centuries is an example. More recently Michael Denton has been exploring laws of form in the context of protein folding. Even a non-Darwinian form of selection and variation can accommodate teleology, provided variations are under intelligent control (perhaps at the level of quantum events) or the environment is carefully fine-tuned by an intelligence to select for appropriate variations. (Darwin's contemporary Asa Gray held to a teleological form of variation.) And then there are the more frankly vitalistic options, like Aristotelian entelechy and the Stoic world-soul, and more recently morphogenetic fields (as in the work of Rupert Sheldrake). All these options, and others as well, are compatible with intelligent design. Intelligent design's only concern is that secondary causes leave room for teleology and that this teleology be empirically detectable.

Nevertheless, a design-theoretic view of evolution would be very different from evolution as it is now conceived. Evolution, as currently presented in biology textbooks, operates not just by secondary causes but by material mechanisms. Many scientists make the mistake of conflating the two. While material mechanisms are perfectly acceptable secondary causes, secondary causes need not be material mechanisms. For instance, although we have every reason to think that Leonardo da Vinci, in painting the Mona Lisa, acted as a secondary cause (we know of no miracles or divine interventions responsible for that painting), we have no reason to think that he acted as a material mechanism (i.e., as an automatic mechanical process simply cranking out that painting by rote). It is a hugely un-

warranted assumption to identify secondary causes and material mechanisms, and yet many scientists and philosophers do just that. The reason
is easy to see: material mechanisms allow for a reductive science in which
the complex can always be explained in terms of the simple. This is convenient as far as it goes. The problem is that in evolutionary biology it
doesn't go very far.

In his *Baltimore Lectures,* Lord Kelvin summed up the attraction of material mechanisms thus: "I never satisfy myself until I can make a mechanical model of a thing. If I can make a mechanical model I can understand
it. As long as I cannot make a mechanical model all the way through, I cannot understand." A mechanism is a well-defined process where each step
of the process leads predictably to the next. A mechanism can be deterministic, in which case one step leads with certainty to the next. Or it can be
stochastic, in which case one step leads with a given probability to the
next. Preeminent among stochastic mechanisms is, of course, the Darwinian mechanism of natural selection and random variation.

Naturalistic forms of evolution depend exclusively on such nonpurpo-
sive material mechanisms. Thus for a naturalistic evolutionary process,
the origin of any species gives no evidence of actual design because mindless material mechanisms do all the work. Within a teleological form of
evolution, by contrast, the origin of some species and biological structures
could give evidence of actual design and demonstrate the inadequacy of
material mechanisms to do such design work. Thus naturalistic evolution
and teleological evolution could have different empirical content and be
distinct scientific theories. Indeed, there are forms of teleological evolution
that are entirely compatible with intelligent design and that involve no
break in secondary causes.

To summarize, the crucial question for intelligent design is not how organisms emerged (e.g., by gradual evolution or sudden special creation)
but whether a designing intelligence made a discernible difference—
regardless of how they emerged. As a consequence, intelligent design has
no stake in upsetting secondary causation. It seeks only to expose naturalism's fallacious attempt to conflate secondary causes with material mechanisms. Accordingly, design can be real and discernible without requiring
an explicit "design event," like a special creation, miracle or supernatural
intervention. At the same time, for evolutionary change to exhibit actual
design would mean that material mechanisms by themselves were inadequate to produce that change.

The question, then, that requires investigation is not simply what are the limits of evolutionary change but what are the limits of evolutionary change when that change is limited to material mechanisms. This in turn requires examining the material factors within organisms (change factors) and in their environments (environmental factors) capable of effecting evolutionary change. The best evidence to date indicates that these factors are inadequate to drive full-scale macroevolution. Something else is required—intelligence.

24

MIRACLES AND
COUNTERFACTUAL SUBSTITUTION

Does intelligent design require miracles?
And if so, wouldn't that place it outside the
bounds of science?

WITH THE RISE OF MODERN SCIENCE, miracles have come to denote a violation or suspension or overriding of natural laws. Yet the classical conception of miracles is quite different. The word *miracle* comes from the Latin and refers to something that inspires wonder or amazement. (The corresponding term in the Greek New Testament is *thauma*, which means the same thing.) Accordingly, a miracle is something that inspires wonder. And since commonplace things don't inspire wonder, miracles are rare or unusual events. We might therefore say that within the classical conception, a miracle is a highly improbable event. To be sure, there is more to the concept, but there's at least that much. Note that this conception introduces no metaphysical presuppositions. In particular, it doesn't prejudge whether nature is ruled by inviolable laws or whether God or some supernatural agent would be violating those laws to perform a miracle. To define a *miracle* as a violation or suspension or overriding of natural laws is already to presuppose what nature is like (namely, that nature is a closed causal nexus governed by inviolable rules). It is also to impose prior limits on divine action.

Even so, suppose for the sake of argument that we go with the view of miracles given to us by modern science. To call something a miracle, then, is to say that a natural cause was all set to make one thing happen but instead something else happened. In other words, miracles involve a counterfactual substitution. Although the term *counterfactual substitution* is recent, the idea is ancient and was explicitly described in counterfactual terms by the theologian Friedrich Schleiermacher. The idea is that natural processes are ready to make outcome X occur, but instead outcome Y occurs. Thus, for instance, with the body of Jesus dead and buried in a tomb for three days, natural processes are ready to keep that corpse a corpse (i.e., the outcome X). But instead that body resurrects (i.e., the miraculous outcome Y).

Does design require miracles in this counterfactual sense? I submit that it does not. To see this, consider that when humans act as intelligent agents, there is no reason to think that any natural law is broken. Likewise, should an unembodied designer act to bring about a bacterial flagellum, there is no reason on its face to suppose that this designer did not act consistently with natural laws. It is, for instance, a logical possibility that the design in the bacterial flagellum was front-loaded into the universe at the big bang and subsequently expressed itself in the course of natural history as a miniature motor-driven propeller on the back of the *E. Coli* bacterium. Whether this is what actually happened is another question, but it involves no contradiction of natural laws and gets around the usual charge of miracles.

Design critic Howard Van Till disagrees. He rejects my claim that intelligent design, by detecting design in nature and especially in biological systems, doesn't require counterfactual substitution. Instead, he argues that intelligent design is fundamentally committed to miracles in the counterfactual substitution sense:

> How could the Intelligent Designer bring about a *naturally impossible outcome* by interacting with a bacterium in the course of time without either a suspension or overriding of natural laws? Natural laws were set to bring about the outcome, no flagellum. Instead, a flagellum appeared as the outcome of the Intelligent Designer's action. Is that not a miracle, even by Dembski's own definition? How can this be anything other than a *supernatural intervention?*

(See his review of my book *No Free Lunch* at <www.aaas.org/spp/dser/evolution/perspectives/vantillecoli.pdf>.)

Van Till's argument hinges on how to interpret the expression "naturally impossible outcome." To see what's at stake, imagine throwing a bunch of Scrabble pieces and seeing them spell out Hamlet's soliloquy. Is this a naturally impossible outcome? It certainly is highly improbable, and such improbability often leads us to attribute impossibility (a pragmatic sort of impossibility). But would such a wildly improbable event require a miracle in the counterfactual-substitution sense of impossibility? Not at all. Scrabble pieces thrown at random are not, as Van Till might put it, "set to bring about the outcome, no Hamlet's soliloquy." Randomness, by definition, has free access to the entire reference class of possibilities that is being sampled. Any possibility from the reference class is therefore fair game for the random process—in this case, the random throwing of Scrabble pieces. It's therefore not the case that this random process was set to bring about "no Hamlet's soliloquy."

Similar considerations apply to the bacterial flagellum. It's not that nature was conspiring to prevent the flagellum's emergence and that a designer was needed to overcome nature's inherent preference for some other outcome (as with counterfactual substitution). Rather, the problem is that nature had too many options and without design couldn't sort through all those options. It's not that natural laws are set to bring about the outcome of no flagellum. The problem is that natural laws are too unspecific to determine any particular outcome. That's the rub. Natural laws are compatible with the formation of the flagellum but also compatible with the formation of a plethora of other molecular assemblages, most of which have no biological significance.

To return to the Scrabble analogy, there's nothing in the throwing of Scrabble pieces that prevents them from spelling Hamlet's soliloquy. This is not like releasing a massive object in a gravitational field that, in the absence of other forces, must move in a prescribed path. For the object to move in any other path would thus entail a counterfactual substitution and therefore a miracle. But with the Scrabble pieces, there is no prescribed arrangement that they must assume. Nature allows them full freedom of arrangement. Yet it's precisely this freedom that makes nature unable to account for specified outcomes of small probability. Nature, in this case, rather than being intent on doing only one thing, is open to doing any number of things. Yet when one of those things is a highly improbable specified event (be it spelling out Hamlet's soliloquy with Scrabble pieces or forming a bacterial flagellum), design becomes the required inference. Van Till has therefore missed the point: it is not counterfactual substitution (and therefore not miracles) but the incompleteness of natural processes that the design inference uncovers.

It follows that intelligence can act without miraculous intervention, in concert with material mechanisms, and yet without being reducible to material mechanisms. For a miracle to take place, nature must get diverted from its ordinary course by doing something other than nature was prepared to do. But nothing like this is required for intelligent design. No physical process need be violated for nature to display the effects of intelligence. With Scrabble pieces taken individually, their motion may be entirely in accord with known natural laws. And yet the coordinated motion of all these pieces might end in a meaningful sentence that must be attributed to intelligence. Natural forces and intelligent agency can act together without one violating the other. Thus intelligent design requires no miracles in the counterfactual substitution sense.

But suppose for the sake of argument that a designing intelligence responsible for biological complexity is in fact a miracle worker that *(presto chango!)* puts biological design into the natural world by means of miracles—miracles in the counterfactual substitution sense of violating or suspending or overriding natural laws. Would that put the design of biological systems outside science? I submit that it would not. Granted, miracles, by replacing an event that ordinarily would have happened with one that ordinarily would not, would entail a gap in the chain of natural causes. But any such gap would itself be open to scientific scrutiny, as would the events on either side of the gap. What's more, if the event that constitutes a miracle (i.e., a counterfactual substitution) clearly exhibits marks of having been intelligently designed, it could issue in a design inference that convincingly implicates design even though its precise causal antecedents may be unclear.

For the convinced naturalist, such a liberal policy toward miracles in science is anathema. The problem is that on the assumption of naturalism any event purported to be a miracle requires causal antecedents that via unbroken natural laws were responsible for it. The causal antecedents and natural laws involved in the event's production are presumed to be there even if we have yet to identify them. This is the standard way of dissolving an event's claim to fame as a miracle. Consequently, the event is no longer a miracle and there is no causal gap between it and its causal antecedents. Any causal gap between the supposed miracle and its causal antecedents merely reflects an incomplete or inadequate characterization by scientists of those antecedents. When these are properly understood, the gap disappears.

For the naturalist, then, there are no miracles and indeed can be no miracles. Any gaps reside not in nature but merely in our minds and reflect the deficient state of our current scientific knowledge. But how is the truth of that claim to be decided? If we allow that miracles (in the sense of causal discontinuities or counterfactual substitutions) might occur and be subject to scientific inquiry, then one could as a matter of empirical inquiry reject miracles by arguing that no good evidence for them obtains. But if one simply by naturalistic fiat pronounces that miracles cannot happen, then one has made an a priori judgment about a question that properly belongs to scientific investigation. It is a fully scientific question whether the laws of nature are complete and properly characterize everything that occurs in nature (making all gaps reside not in nature but solely in our intellects). The flipside is therefore necessarily also a fully scientific question, namely, whether there are ontological gaps in nature that no amount of fiddling with material mechanisms and the laws that characterize them will ever bridge. By artificially closing the door to such questions, naturalism stipulates the way nature must be apart from any empirical investigation. This is armchair philosophy and not science.

25

THE SUPERNATURAL

Isn't the designer to which intelligent design attributes biological complexity a supernatural agent and therefore outside the bounds of science?

I'VE NEVER LIKED THE TERM *SUPERNATURAL*. The problem with terms like *supernatural* and *supernaturalism* (and I include here Howard Van Till's variant of *extra-natural assembly*) is that they tacitly presuppose that nature is the fundamental reality and that nature is far less problematic conceptually than anything outside or beyond nature. The *super* in *supernatural* thus has the effect of a negation.

But what if nature is itself a negation or reaction against something else? For the theist (though not for the panentheist of process theology, much less for the typical naturalist), nature is not a self-subsisting entity but an entirely free act of God. Nature thus becomes a derivative aspect of ultimate reality—an aspect of God's creation, and not even the whole of God's creation at that. (Theists typically ascribe to God the creation of an invisible world that is inhabited by, among other things, angels.) Hence, for the theist attempting to understand nature, God as creator is fundamental, the creation is derivative, and nature as the physical part of creation is still further downstream.

Now, from the vantage of intelligent design, treated strictly as a scientific inquiry, no theological or antitheological position has a privileged place. Intelligent design, as a scientific research program, attempts to determine whether certain features of the natural world exhibit signs of having been designed by an intelligence. This intelligence could be E.T. or a telic principle immanent in nature or a transcendent personal agent. These are all, at least initially, live options. The problem with E.T., of course, is that it implies a regress—where did E.T. come from? The same question doesn't apply, at least not in the same way, to telic principles or to transcendent personal agents because the terms of the explanation are different. E.T. is an embodied intelligence, and that embodiment itself needs explanation. Telic principles and transcendent agents are unembodied. That raises its own issues, but they are a different set of issues.

Suppose now that research in intelligent design discovers an intelligence or multiple intelligences that cannot be confined to space, time, matter and energy. Forget about the term *supernatural* and the presuppositional baggage it carries. What if the designing intelligence(s) responsible for biological complexity cannot be confined to physical objects? Why should that burst the bounds of science? Certainly science has the physical world as its proper object of study. Science studies and tries to explain the things that are happening in the physical world. But if studying the physical world is the sole criterion for determining whether an explanation is properly scientific, then design must constitute a part of science, for the specified complexity that intelligent design studies is instantiated throughout the physical world, notably in living things, and these form an integral part of it. When design is faulted for not properly being a part of science, however, it is not for making living things an object of study. Rather, it is for attributing living things to nonnaturalistic causes—to supernatural designers—and thereby making these nonnaturalistic causes objects of study as well.

In answering this criticism, let us first of all be clear that intelligent design does not require miracles in the sense of violations of natural law. Just as humans do not perform miracles every time they act as intelligent agents, so too there is no reason to assume that for a designer to act as an intelligent agent requires a violation of natural laws. There's an important contrast to keep in mind here. Science, so most naturalists contend, studies natural causes whereas to introduce design is to invoke supernatural causes. Thus intelligent design, by invoking the supernatural, is said to burst the bounds of science.

But the contrast between natural and supernatural causes is the wrong contrast. The proper contrast is between *undirected natural causes* on the one hand and *intelligent causes* on the other. Intelligent causes can work with natural causes and help them to accomplish things that undirected natural causes cannot. Undirected natural causes can explain how ink gets applied to paper to form a random inkblot but cannot explain an arrangement of ink on paper that spells a meaningful message. To obtain such a meaningful arrangement requires an intelligent cause. Whether an intelligent cause is located within or outside nature is a separate question from whether an intelligent cause has acted within nature. Design has no prior commitment against naturalism or for supernaturalism. Consequently, science can offer no principled grounds for excluding design or relegating it to the sphere of religion.

Given that intelligent design has no prior commitment to supernaturalism (and indeed regards supernaturalism as a red herring), what then remains of the criticism that science must be confined exclusively to the natural or physical? Implicit in this criticism is a theory of reference specifying the types of entities to which a scientific theory may legitimately refer. Accordingly, scientific terms must refer to entities strictly locatable in space and time. Given such a theory of reference, the mere fact that a designer might not be physically embodied and therefore locatable in space and time is enough to make explanations that invoke design nonnaturalistic and therefore nonscientific. Even those unobservable theoretical entities of physics like fields, potentials, quarks and strings are at least in principle locatable in space and time. But because a designer need not be, designers must be excluded from science.

This criticism miscarries. For one thing, it presupposes a realist view of scientific explanation. If, for instance, one takes an antirealist view of scientific explanation, then the spatiotemporal location of the entities posited by a scientific theory becomes moot because space and time themselves become conceptual constructs. On an antirealist view, the important thing is not where those entities are located, since scientific theories describe not so much what is "out there" as they do our way of conceptually structuring the world. Thus, the important thing on the antirealist view is whether the entities we posit are empirically adequate and conceptually fruitful and whether they possess superior explanatory power.

If a theory of design in biology should prove successful, would it follow that the designer posited by this theory is real? An antirealist about science could simply regard the designer as a regulative principle—a conceptually useful device for making sense out of certain facts of biology—without assigning the designer any weight in reality. Ludwig Wittgenstein regarded the theories of Copernicus and Darwin not as true but as fertile points of view. One's first interest as a scientist working on a theory of intelligent design is whether design provides powerful new insights and fruitful avenues of research (see chapters forty-three and forty-four). The metaphysics underlying such a theory, and in particular the ontological status of the designer, can then be taken up by philosophy and theology. Indeed, one's metaphysics ought to be a matter of indifference to one's scientific theorizing about design.

But there's a more important reason why science must remain open to designers lacking spatiotemporal characteristics: the only reason for requiring science to ignore entities not locatable in space and time is that we know in advance that such entities do not exist or, if they do exist, that they can have no conceivable relevance to what happens in the world. Do such entities exist? Can they have empirical consequences? Are they relevant to what happens in the world? Such questions cannot be prejudged except on metaphysical grounds. To prejudge these questions is therefore to make certain metaphysical commitments about what there is and what can influence events in the world. Such commitments are gratuitous to the practice of science. Intelligent design puts the biological evidence for design under the microscope. The fact that the designing intelligence responsible for life can't be put under the microscope poses no obstacle to science. We learn of this intelligence as we learn of any other intelligence—not by studying it directly but through its effects.

26

EMBODIED AND UNEMBODIED DESIGNERS

Would the design produced by an unembodied designer be accessible to scientific investigation in the same way as the design produced by an embodied designer?

ALL OF US HAVE IDENTIFIED THE EFFECTS OF EMBODIED DESIGNERS. Indeed, fellow human beings constitute our best example of such designers. We are also confident that we could identify the effects of embodied designers that are not animal or human, like extraterrestrial intelligences (such as those in the movies *Contact* and *2001: A Space Odyssey*). But what about unembodied designers? With respect to intelligent design in biology, for instance, Elliott Sober wants to know what sorts of biological systems should be expected from an unembodied designer. Sober claims that if the design theorist cannot answer this question, then intelligent design is untestable and therefore unfruitful for science.

Yet to place this demand on design hypotheses is ill-conceived. We infer design regularly and reliably without necessarily knowing the characteristics of the designer or being able to assess what the designer is likely to do. Although it's an interesting question whether a designer is embodied, a designer's embodiment is of no evidential significance for determining whether something was designed in the first place. The reason embodiment is irrelevant to detecting design is that we always attribute design on the basis of an inference from empirical data and never on the basis of a direct encounter with a designer's mental processes.

We do not get into the mind of designers and thereby attribute design. Rather, we look at effects in the physical world that exhibit clear marks of intelligence and from those marks infer a designing intelligence. This is true even for those most uncontroversial of embodied designers, namely, our fellow human beings. We recognize their intelligence not by merging with their minds but by examining their actions and determining whether those actions display marks of intelligence. A human being who speaks coherent sentences displays intelligence. A human being who continuously mumbles the same nonsense syllable displays no intelligence and provides no justification for ascribing it.

What about cases where a clear example of design can be plausibly attributed only to an unembodied designer? Darwinian naturalists pretend such design is undetectable simply because we can't reduce the actions of the designing agent to purely natural causes. Robert Pennock, for instance, requires that design inferences be based on "known types of causal processes." He means that the underlying causal processes must be fully reducible to material mechanisms before a design inference may legitimately be drawn. But that is precisely the point at issue, namely, whether intelligent agency reduces to or transcends material mechanisms. Specified complexity, as a criterion for detecting design, allows that question to be assessed without prejudice. Pennock, on the other hand, by presupposing naturalism, has stacked the deck so that only one answer is possible.

Larry Arnhart is another critic who remains unconvinced that a design inference can validly infer an unembodied intelligence. Arnhart maintains that our knowledge of design arises in the first instance not from any inference but from introspection of our own human intelligence. As a consequence, he concludes that we have no empirical basis for inferring design whose source is unembodied. Though at first blush plausible, this argument quickly collapses when probed. Jean Piaget, for instance, would have rejected it on developmental grounds: babies do not make sense of intelligence by introspecting their own intelligence but by coming to terms with the effects of intelligence in their external environment. For example, they see the ball in front of them and then see it taken away, and they learn that Daddy is moving the ball—thus reasoning from effect to intelligence. Introspection plays at best a secondary role in how we initially make sense of intelligence and design.

Even later in life, when we have attained full self-consciousness and when introspection can be performed with varying degrees of reliability, design is still inferred. Indeed, introspection must always remain an inadequate method of assessing intelligence and attributing design. By definition, intelligence presupposes the power or facility to choose between options, which coincides with the Latin etymology of *intelligence*, namely, "to choose between." Introspection is therefore entirely the wrong instrument for assessing intelligence. (It can even lead to delusions about our intelligence.) For instance, I cannot by introspection assess my intelligence as a carpenter. I actually have to get out and build a deck or cabinet or table. How I perform the complicated sequence of decisions to cut here, saw there and nail this to that—and not any act of introspection—will determine whether and to what degree intelligence can be attributed to my carpentry. The only way to assess intelligence is to test it and see what it does. (That's why education proceeds by tests and examinations rather than by introspective reports where students profess, or perhaps protest, their competence.) What's more, the primary, empirically verifiable thing that intelligences do is generate specified complexity.

I therefore continue to maintain that intelligence is always inferred, that we infer it through well-established methods and that there is no principled way to argue that the work of embodied designers is detectable whereas the work of unembodied designers isn't. This is the rub. And this is why intelligent design is such an intriguing intellectual possibility: it threatens to bring the ultimate questions down to earth. Convinced Darwinists like Pennock and Arnhart therefore need to block the design inference whenever it threatens to implicate an unembodied designer. Embodied designers are okay. That's why Francis Crick can get away with his directed panspermia theory in which intelligent aliens (who are embodied designers) travel by spaceships to planet earth and seed it with life from outer space.

So long as the designer is embodied, Darwinists can claim that the designer is an evolved intelligence that arose via the Darwinian mechanism. Unembodied designers, however, are strictly outside the bounds of the Darwinian mechanism and therefore strictly proscribed. Back when Darwinists felt sure that primitive life could have sprung from nonlife on the early earth, any such theory of alien life seeding the earth would have been rightly ridiculed as so much bad science fiction. But as soon as we learned enough about the astonishing complexity of the cell and the unfeasibility of such life spontaneously emerging on the early earth, the extraterrestrial Johnny Appleseed scheme suddenly became sound and cogent science. Why? Because it saves the dogmatic naturalists from fairly and honestly considering the evidence for an unembodied designer.

Not only is there no evidential significance as to whether a designer is embodied or unembodied for drawing a design inference, but refusing to countenance the possibility of unembodied designers impedes scientific inquiry. To see this, consider the following variant of the Explanatory Filter. I call it the Naturalized Explanatory Filter (see p. 196). Since the most typical candidate for an unembodied designer is God, I have formulated the Naturalized Explanatory Filter with explicit reference to God. Nevertheless, one is free to substitute in place of God any unembodied entity that is unacceptable on naturalistic grounds and yet implicated by a design inference. The Naturalized Explanatory Filter accurately captures how scientific naturalism tries to account for the design of natural systems for which no embodied designer can plausibly be invoked as an explanation.

Like the Explanatory Filter discussed in chapter eleven, the Naturalized Explanatory Filter assesses contingency, complexity and specification. Moreover, so long as no naturalistically unacceptable entity (like God) is implicated, the filter properly sorts through the three primary modes of explanation: necessity, chance and design. Nonetheless, as soon as a naturalistically unacceptable entity is implicated, the Naturalized Explanatory Filter—instead of coming to terms with it and squarely acknowledging that there is now a design problem here—conveniently adds a fourth decision node that cycles the explanatory analysis back to the beginning of the flowchart and thereby ensures that only necessity and chance will receive further consideration. But since design is in fact implicated (as would be clear from applying the unrigged Explanatory Filter), the explanatory analysis can never terminate. The Naturalized Explanatory Filter therefore artificially forces suspension of judgment whenever designers unacceptable to naturalism are implicated.

It follows that even if an unembodied intelligence is responsible for the design displayed in some phenomenon, a science committed to the Naturalized Explanatory Filter will be sure never to discover it. A science that on a priori grounds refuses to consider the possibility of unembodied designers artificially limits what it can discover. Instead of rejecting design as the conclusion of a sound scientific argument, this approach stipulates it out of existence. Essential to science is a spirit of free and open inquiry. Scientific naturalism, and the Naturalized Explanatory Filter that it endorses, violates that spirit. Because detecting design is a perfectly ordinary, rational activity, science needs a rigorous, even-handed design-detection filter. The Naturalized Explanatory Filter makes clear how naturalism has been used historically to derail the design inference whenever design has come too close to challenging naturalism. It goes without saying that the Naturalized Explanatory Filter is a subterfuge.

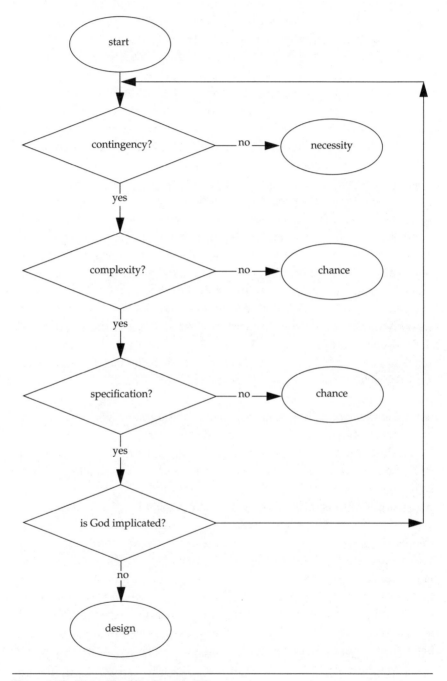

Figure 4. The Naturalized Explanatory Filter

27

THE DESIGNER REGRESS

*If nature exhibits design, who or what
designed the designer?*

ACCORDING TO RICHARD DAWKINS (in *The Blind Watchmaker*), to explain
by means of an unembodied designer is

> to explain precisely nothing, for it leaves unexplained the origin of
> the Designer. You have to say something like "God was always
> there," and if you allow yourself that kind of lazy way out, you might
> as well just say "DNA was always there," or "Life was always there,"
> and be done with it.

Dawkins takes this line because he, like many scientists and philosophers,
is convinced that proper scientific explanations must be reductive—
explaining the complex in terms of the simple. Thus Dawkins adds, "The
one thing that makes evolution such a neat theory is that it explains how
organized complexity can arise out of primeval simplicity."

Dawkins equates proper scientific explanation with what he calls "hier-
archical reductionism," according to which "a complex entity at any particular level in the hierarchy of organization" must properly be explained
"in terms of entities only one level down the hierarchy." Thus, evolved designers are okay because ultimately they promise to submit to reductive
explanations. (Dawkins is a universal Darwinist, so any designers anywhere in the universe must have evolved via the Darwinian mechanism.)
On the other hand, unevolved designers are not okay because they can
never submit to reductive explanations.

In responding to the who-designed-the-designer question, it is therefore best first to dispense with Dawkins's reductionist view of science.
This is easily done. While no one will deny that reductive explanation is
extremely effective within science, it is hardly the only type of explanation
available to science. The divide-and-conquer mode of analysis behind re-

ductive explanation has strictly limited applicability within science. Complex systems theory has long rejected a reductive bottom-up approach to complex systems. To understand complex systems properly requires a top-down approach that focuses on global relationships between parts as opposed to analysis into individual parts. So too, intelligent design is an integrative, top-down theory of complex structures. Integration is as much a part of science as are reduction and analysis.

The regress implicit in the who-designed-the-designer question is no worse here than elsewhere in science. Such regresses arise whenever scientists introduce novel theoretical entities. For instance, when Ludwig Boltzmann introduced his kinetic theory of heat back in the late 1800s and invoked the motion of unobservable particles (what we now call atoms and molecules) to explain heat, one might just as well have argued that such unobservable particles do not explain anything because they themselves need to be explained.

It is always possible to ask for further explanation. Nevertheless, at some point scientists stop and content themselves with the progress they've made. Boltzmann's kinetic theory explained things that the old phenomenological approaches to heat failed to explain—for instance, why shaking a container filled with a gas caused the temperature of the gas to increase. Whereas the old phenomenological approach provided no answer, Boltzmann's kinetic theory did: shaking the container caused the unobservable particles making up the gas to move more quickly and thus caused the temperature to rise.

So too with design, the question is not whether design theorists have resolved all lingering questions about the designing intelligence responsible for specified complexity in nature. Such questions will always remain. Rather, the question is whether design does useful conceptual work—a question that Dawkins's criticism leaves unanswered. Design theorists argue that intelligent design is a fruitful scientific theory for understanding systems like Michael Behe's irreducibly complex biochemical machines. Such an argument has to be taken on its own merits. It is a scientific argument.

Intelligent design as a scientific research program looks for empirical markers of intelligence in nature. Irreducible complexity is one such marker in biology, and it reliably points to an intelligence. Now the designer who is behind that design is, as far as we can tell, not part of nature (at least as nature is now understood by the scientific community). Consequently there is no "marker" attached to this designer indicating that this designer is in turn designed. The theory of intelligent design therefore avoids the "design regress" in which we must—to stay consistent with our methods of design detection—answer whether the designer is designed. The designer is not an event, object or structure. Consequently, the designer, though capable of producing phenomena that exhibit empirical marks of intelligence, cannot as such exhibit any empirical marks of intelligence. The question whether the designer is designed simply does not arise within a scientific theory of design.

This has nothing to do with intelligent design evading difficult or embarrassing problems. Design-theoretic explanations are proximal or local explanations rather than ultimate explanations. Design-theoretic explanations are concerned with determining whether some particular event, object or structure exhibits clear marks of intelligence and can thus be legitimately ascribed to design. Consequently, design-theoretic reasoning does not require the who-designed-the-designer question to be answered for a design inference to be valid. There is explanatory value in attributing the Jupiter Symphony to the artistry (design) of Mozart, and that explanation suffers nothing by not knowing who designed Mozart. Likewise in biology, design inferences are not invalidated for failing to answer Dawkins's who-designed-the-designer question.

The who-designed-the-designer question is best interpreted as a metaphysical rather than a scientific question. As such, it is a call for ultimate rather than proximal explanations. Proximal explanations are contextual and local, focusing on particular features of the world at particular times and places. Ultimate explanations, on the other hand, are global and encompassing, focusing on the entire world across time and looking to a final resting place of explanation.

The naturalist is likely to posit Nature (writ large) or the Universe (also writ large) or mass-energy or superstrings or some such entity as the final resting place for explanation. Likewise, the design theorist is likely to posit a generic designer or specified complexity or immanent teleology or God as the final resting place of explanation. This does not mean that all ultimate explanations are equivalent (though judging their merits goes beyond the scope of science). Nor does this mean that science is irrelevant to deciding among ultimate explanations. Darwinism conduces toward naturalism whereas intelligent design, at least in contemporary Western culture, conduces toward theism. The crucial point, however, is that design-theoretic explanations can be legitimate without being ultimate, and that is especially true in biology.

28

SELECTIVE SKEPTICISM

Why is professional skepticism so
antagonistic toward intelligent design?
What are skepticism's prospects for
unseating intelligent design?

PROFESSIONAL SKEPTICS MAKE A BUSINESS OF keeping science free from superstition. Many of them now view intelligent design as the most serious threat to science on the cultural landscape. They worry that intelligent design will subvert science—a worry that became particularly evident at a special forum on intelligent design that they invited me to address. That forum took place at the Fourth World Skeptics Conference in Burbank, California, on June 21, 2002. The aim of that conference was to chart skepticism's prospects over the next quarter century. At the conference, intelligent design emerged as skepticism's new primary villain.

Skepticism has a standard bag of tricks for keeping a gullible public in line. I want to lay out here why that bag of tricks is unlikely to succeed against intelligent design. One problem facing skepticism is intelligent design's inherent reasonableness and growing respectability. A few years ago skeptic Michael Shermer wrote a book titled *Why People Believe Weird Things.* Most of the weird things Shermer discusses in that book are definitely on the fringes, like holocaust denial, alien encounters and witch crazes—hardly the sort of stuff that's going to make it into the public school science curriculum. Intelligent design, by contrast, is becoming thoroughly mainstream and threatening to do just that.

Gallup poll after Gallup poll confirms that about 90 percent of the U.S. population believes that some sort of design is behind the world. At the time of this writing, Ohio is the epicenter of the evolution-intelligent design controversy. Recent polls conducted by the *Cleveland Plain Dealer* (June 9, 2002) found that 59 percent of Ohioans want both evolution and intelligent design taught in their public schools. Another 8 percent want only intelligent design taught. And another 15 percent do not want the teaching of intelligent design mandated but do want to allow evidence against evolution to be presented in public schools. The arithmetic is clear.

Perhaps the most telling finding of this poll is how Ohioans view the consequences for their state of having intelligent design taught in their public schools. According to the *Cleveland Plain Dealer* (June 9, 2002), "About three of every four respondents said including intelligent design in the curriculum would have either a positive effect or no effect on the state's reputation or its ability to attract new business." One could hardly imagine the same response if the question were whether to teach astrology, witchcraft or flat-earth geology. Intelligent design has already become mainstream with the public at large.

Even so, the mainstreaming of intelligent design doesn't cut any ice with skeptics. Skeptics know all about logical and informal fallacies, and the *argumentum ad populum* (appeal to the masses) heads the list. Skepticism purports to keep a gullible public honest. Accordingly, just because intelligent design is acceptable to most Americans doesn't mean that it deserves acceptance. (Witness America's fascination with horoscopes.) All the same, there's reason to think that the usual skeptical assaults are not going to prosper against intelligent design.

One of skepticism's patron saints, H. L. Mencken, remarked, "For every problem, there is a neat, simple solution, and it is always wrong." Yet in writing about Darwin's theory, Stephen Jay Gould remarked, "No great theory ever boasted such a simple structure" (quoted from Gould's introduction to Carl Zimmer's *Evolution: The Triumph of an Idea*). Intelligent design claims that Mencken's insight applies to evolutionary biology, overturning not just mechanistic accounts of evolution but skepticism itself.

Skepticism, to be true to its principles, must be willing to turn the light of scrutiny on anything. And yet that is precisely what it cannot afford to do in the controversy over evolution and intelligent design. The problem with skepticism is that it is not a pure skepticism. Rather, it is a selective skepticism that desires a neat and sanitized world which science can in principle fully characterize in terms of unbroken natural laws.

Indeed, why is the premier skeptic organization in the world known as CSICOP, the Committee for the Scientific Investigation of Claims of the Paranormal? (They sponsored the conference mentioned above.) Though I'm assured that the "COP" in CSICOP is purely coincidental, it is nonetheless singularly fitting. CSICOP is in the business of policing claims about the paranormal. The paranormal, by being other than normal, threatens the tidy world issuing from skepticism's materialistic conception of science.

No other conception of science will do for skepticism. The normal is what is describable by a materialistic science. The paranormal is what's not. Given the skeptic's faith that everything is ultimately normal, any claims about the paranormal must ultimately be bogus. And since intelligent design claims that an intelligence not ultimately reducible to material mechanisms might be responsible for the world and various things we find in the world (not least ourselves), it too is guilty of transgressing the normal and must be relegated to the paranormal.

There is an irony here. The skeptic's world, in which intelligence is not fundamental and the world is not designed, is a rational world because it proceeds by unbroken natural law: cause precedes effect with inviolable regularity. In short, everything proceeds "normally." On the other hand, the design theorist's world, in which intelligence is fundamental and the world is designed, is not a rational world because intelligence can do things that are unexpected. In short, it is a world in which some things proceed "paranormally."

To allow an unevolved intelligence a place in the world is, according to skepticism, to send the world into a tailspin. It is to exchange unbroken natural law for caprice and thereby to destroy science. And yet it is only by means of our intelligence that science is possible and that we understand the world. Thus, for the skeptic, the world is intelligible only if it starts off without intelligence and then evolves intelligence. If it starts out with intelligence and evolves intelligence because of a prior intelligence, then, for the skeptic, the world becomes unintelligible.

The logic here is flawed, but once in its grip, there is no way to escape its momentum. That is why evolution is a nonnegotiable for skepticism. For instance, on two occasions I offered to join the editorial advisory board of Michael Shermer's *Skeptic* magazine to be its resident skeptic regarding evolution. Though Shermer and I are quite friendly, he never took me up on my offer. Indeed, he can't afford to. To do so is to allow that an intelligence outside the world might have influence in the world. That would destroy the world's autonomy and render effectively impossible the global rejection of the paranormal that skepticism requires. It's no accident that the photo of Shermer which appears in his books shows him smiling with a bust of Darwin and a collection of writings by or about Darwin behind him.

Skepticism therefore faces a curious tension. On the one hand, to maintain credibility it must be willing to shine the light of scrutiny everywhere and thus, in principle, even on evolution. On the other hand, to be the scourge with which to destroy superstition and whip a gullible public into line, it must commit itself to a materialistic conception of science and thus cannot afford to question evolution. Intelligent design exploits this tension and thereby turns the tables on skepticism.

What, then, are skepticism's prospects for unseating intelligent design? To answer this question, let's review what intelligent design has going for it:

1. A method for design detection. There's much discussion about the validity of specified complexity as a method for design detection, but judging by the response it has elicited over the last five years, this method is not going away. Some scholars (such as Elliott Sober) think it merely codifies an argument from ignorance. Others (such as Paul Davies) think that it's onto something important. The point is that there are major players who are not intelligent design proponents who disagree. Such disagreement indicates that issues of real intellectual merit need to be decided and that we're not dealing with a crank theory.

2. Irreducibly complex biochemical systems. There exist systems like the bacterial flagellum. These exhibit specified complexity. Moreover, the biological community does not have a clue how they emerged by material mechanisms. The great promise of Darwinian and other naturalistic accounts of evolution was to show how known material mechanisms operating in known ways could produce all of biological complexity. That promise is now increasingly recognized as unfulfilled and even unfulfillable. Franklin Harold (who is not a design proponent), in his most recent book for Oxford University Press, *The Way of the Cell*, states, "There are presently no detailed Darwinian accounts of the evolution of any biochemical or cellular system, only a variety of wishful speculations." Intelligent design contends that our ignorance here indicates not minor gaps in our knowledge of biological systems that promise readily to submit to tried-and-true mechanistic models but rather vast conceptual lacunae that are bridgeable only by radical ideas like design.

3. Challenge to the status quo. Let's face it, in educated circles Darwinism and other mechanistic accounts of evolution are utterly status quo. That has advantages and disadvantages for its proponents. On the one hand, it means that the full resources of the scientific and educational establishment are behind the evolutionary naturalists, which they can use to squelch dissent and push their agenda. On the other hand, it means that they are in danger of alienating the younger generation—especially to the extent that they are heavy-handed in enforcing materialist orthodoxy (and they've been exceedingly heavy-handed to date)—which thrives on rebellion against the status quo. Intelligent design appeals to the rebelliousness of youth.

4. The disconnect between high and mass culture. The educated elite love mechanistic evolution and the materialist science it helps to underwrite. On the other hand, the masses are by and large convinced of intelligent design. What's more, the masses ultimately hold the purse strings for the educated elite (in the form of educational funding, research funding, scholarships, etc.). This disconnect can be exploited. The advantage that mechanistic evolution has had thus far is providing a theoretical framework, however empirically inadequate, to account for the emergence of biological complexity. The disadvantage facing the intelligent-design-supporting masses is that they've had to rely almost exclusively on pretheoretic design intuitions. Intelligent design offers to replace those pretheoretic intuitions with a rigorous design-theoretic framework that underwrites those intuitions, thus allowing it to go toe-to-toe with standard evolutionary theory.

5. An emerging research community. Intelligent design is attracting bright young scholars who are totally committed to developing intelligent design as a research program. We're still thin on the ground, but the signs I see are very promising indeed. It's not enough merely to detect design, for once it's detected, it must be shown how design leads to fruitful biological insights that could not have been obtained by taking a purely materialist outlook. I'm beginning to see glimmers of such a thriving design-theoretic research program.

What's a skeptic to do against this onslaught, especially when there's a whole political dimension to the debate in which a public tired of being bullied by an intellectual elite finds in intelligent design a tool for liberation? Let me suggest to the skeptic the following action points:

1. Establish the right rhetorical tone. Emphasize science as a great force for enlightenment and contrast it sharply with fanatical religious fundamentalism. Then stress that intelligent design is essentially a religious movement. Generously use the "C-word" to confuse intelligent design with creationism, and emphasize the similarity of creationism with astrology, belief in a flat earth and holocaust denial. Once guilt by association is in place, play on the theme that intelligent design is "deeply flawed" and that the evidence for evolution is "overwhelming." Think of the phrases *deeply flawed* and *overwhelming evidence* not as actual criticisms or arguments but as slogans that evoke the appropriate emotional response. (Compare them to "Don't leave home without it," "This Bud's for you," and "Just do it!") For the record, I own the domain names <www.deeply-flawed.com> and <www.overwhelmingevidence.com> (as well as <www.underwhelmingevidence.com>).

2. Argue for the superfluity of design. This action point is getting increasingly difficult to implement simply on the basis of empirical evidence, but by artificially defining science as an enterprise limited solely to material mechanisms, one conveniently eliminates design from scientific discussion. Thus, any gap in our knowledge of how material mechanisms brought about some biological system does not reflect an absence of material mechanisms in nature to produce the system or a requirement for design to account for the system, but only a gap in our knowledge that's readily filled by carrying on as science has been carrying on.

3. Play the suboptimality card. For most people the designer is a benevolent, wise God. This allows for the exploitation of cognitive dissonance by pointing to cases of incompetent or wicked design in nature. Intelligent design has good answers to this objection, but the problem of evil is wonderfully adept at clouding intellects. This is one place where skepticism does well exploiting emotional responses.

4. Achieve a scientific breakthrough. Provide detailed testable models of how irreducibly complex biochemical systems like the bacterial flagellum could have emerged by material mechanisms. I don't give this possibility much hope, but if skeptics could pull this off, intelligent design would have a lot of backpedaling to do.

5. Paint a more appealing world picture. Skepticism is at heart an austere enterprise. It works by negation. It makes a profession of shooting things down. This doesn't set well with a public that delights in novel possibilities. In his *Art of Persuasion*, Blaise Pascal wrote, "People almost invariably arrive at their beliefs not on the basis of proof but on the basis of what they find attractive." Poll after poll indicates that for most people a mechanistic form of evolution does not provide a compelling vision of life and the world. Providing such a vision is, in my view, skepticism's overriding task if it is to unseat intelligent design. Skeptics have my very best wishes for success in this enterprise.

29

THE PROGRESS OF SCIENCE

Does scientific progress invariably vindicate naturalism and work against intelligent design?

THE MARK OF AN IDEOLOGY IS TRIUMPHALISM, the illusion that success is inevitable. Naturalism is the worst offender I know in this regard. Imagine an unstoppable tidal wave that sweeps away everything in its path. The only recourse is retreat. But no matter to what position one retreats, it is only a matter of time before the tidal wave sweeps away that redoubt as well. In this image, a design-eschewing naturalistic science is the unstoppable tidal wave, and anything not readily assimilated to naturalism (like design, souls, natural law ethics, religion and God) is beating a hasty retreat, always giving up ground and never gaining it back. This is a wonderful image. A lot of people believe it. And it is dead wrong.

Despite all the propaganda to the contrary, science is not a tidal wave that relentlessly pushes back the frontiers of knowledge. Rather, science is an interconnected web of theoretical and factual claims about the world that are constantly being revised and for which changes in one portion of the web can induce radical changes in another. In particular, science regularly confronts the problem of having to retract claims that it once boldly asserted.

Consider the following example from geology. In the nineteenth century the geosynclinal theory was proposed to account for how mountain ranges originate. This theory hypothesized that large troughlike depressions, known as geosynclines, filled with sediment, gradually became unstable, and then, when crushed and heated by the earth, elevated to form mountain ranges. To the question *What happened when?* geologists as late as 1960 confidently asserted that the geosynclinal theory provided the answer. Thus in the 1960 edition of Thomas Clark and Colin Stearn's *Geological Evolution of North America*, the status of the geosynclinal theory was compared favorably with Darwin's theory of natural selection:

The geosynclinal theory is one of the great unifying principles in geology. In many ways its role in geology is similar to that of the theory of evolution, which serves to integrate the many branches of the biological sciences. . . . Just as the doctrine of evolution is universally accepted among biologists, so also the geosynclinal origin of the major mountain systems is an established principle in geology.

Whatever became of the geosynclinal theory? Within ten years of this statement, the theory of plate tectonics, which explained mountain formation through continental drift and sea-floor spreading, had decisively replaced the geosynclinal theory. The history of science is filled with such turnabouts in which confident claims to knowledge suddenly vanish from the scientific literature.

The geosynclinal theory was completely wrong. Thus, when the theory of plate tectonics came along, the geosynclinal theory was overthrown. Often, however, theories are not completely wrong. Instead, they offer some legitimate insights. Nevertheless, upon further investigation, they need to be revised. Frequently, such revision takes the form of a contraction. The problem is that when theories are first proposed, their originators try to push them to account for as much as possible—indeed, for too much. Only later do the limitations of the theory become evident.

It is always a temptation in science to think that one's theory encompasses a far bigger domain than it actually does. This happened with Newtonian mechanics: physicists thought that Newton's laws provided a total account of the constitution and dynamics of the universe. James Maxwell, Albert Einstein and Werner Heisenberg each showed that the proper domain of Newtonian mechanics was far more constricted. (Newtonian mechanics works well for medium-sized objects at medium speeds, but for very fast and very small objects it breaks down, and we need relativity and quantum mechanics, respectively.) So too, the proper domain of the Darwinian selection mechanism is far more constricted than most Darwinists would like to admit. In particular, large-scale evolutionary changes in which organisms gain novel information-rich structures cannot legitimately be derived from the Darwinian selection mechanism.

Sometimes, as in the case of geosynclinal geology, theories are replaced in their entirety by completely new theories. At other times, as with Newtonian mechanics, theories prove inadequate outside a certain range of phenomena and need to be supplemented. (No one any longer learns geosynclinal geology, but all freshman physics students still learn Newtonian mechanics, though later in their course of study they also learn about quantum mechanics and relativity theory.) In both these instances, however, defective theories give way to new and improved theories. But that's not always the case. It's also possible for theories to be overthrown or contracted without offering a replacement theory.

Consider the case of superconductivity. Here science did not require a replacement theory ready and available in order to establish the inadequacy of an existing theory when the experimental evidence went against the existing theory. Such case studies are particularly important in the debate over naturalistic evolution because they show that one may legitimately criticize Darwinism and other naturalistic theories of evolution without having to argue for the adequacy of a replacement theory. Instead of trying to shoehorn recalcitrant data into theories that are empirically inadequate, science is regularly forced to give up overconfident claims that cannot be adequately justified. The rational alternative to Darwin's theory, therefore, need not be intelligent design but could simply be intelligent uncertainty.

The Dutch physicist Kamerling Onnes discovered superconductivity in 1911. Superconductivity refers to the complete disappearance of electrical resistance for materials at low temperatures. Back when Onnes made his discovery, however, there was no theory to account for superconductivity. Such a theory was not proposed until 1957, namely, the BCS theory. This theory is named for John Bardeen, Leon Cooper and John Schrieffer, who received the Nobel Prize in physics for their theory in 1972. The first paragraph of the Nobel press release describes the BCS theory as providing "a complete theoretical explanation of the phenomenon." (See <www.nobel.se/physics/laureates/index.html>.) But the theory didn't stay complete for long. In the 1980s Georg Bednorz and Alexander Müller discovered superconductors at much higher temperatures than previously identified and explained by the BCS theory. To date, no replacement theory for BCS has been found that extends to high-temperature superconductors. BCS, instead of being "*the* theory of superconductivity," now merely explains a quite limited range of superconductors.

Science can get things wrong—indeed, massively wrong. What's more, sometimes we can tell that science has gotten something wrong without having to identify what the correct or true explanation is. Also, unlike religion, science has no prophets. There are no scientific prophets to tell us what course science must take or avoid taking. Different courses need to be tried, and only after they are tried does it become clear what was fruitful and what was fruitless. Criticisms against intelligent design, like the following by Howard Van Till, therefore lie outside science: "The more we learn about the self-organizational and transformational feats that can be accomplished by biotic systems, the less likely it will be that the conditions for [specified complexity] will be satisfied by any biotic system." (See <www.aaas.org/spp/dser/evolution/perspectives/vantillecoli.pdf>.) In other words, the further a materialistic science progresses, the less likely it will be that intelligent design is vindicated.

Van Till is here playing the prophet. There's no unprejudiced reason to think that as our knowledge of natural processes relevant to the formation of biotic systems increases, the improbabilities or complexities associated with such systems will diminish and specified complexity will thereby get refuted or dwindle away. (And since specified complexity is a marker of intelligent design, the detectability of design would thereby also get refuted or dwindle away.) But that's not how probabilities and complexities work. With increasing scientific knowledge, the numbers we calculate may stay the same, diminish or increase (implying that if calculated complexity increases, the challenge of specified complexity will only intensify). This is a special case of a much more general feature of science, namely, that the known is not a reliable guide to the unknown. Indeed, the history of science is a history of surprises.

Van Till's attempt to play the prophet should give us pause. He admits that increasing scientific knowledge might refute an attribution of specified complexity to some biotic system. But if that's a possibility, then certainly it's also a possibility that increasing scientific knowledge might fail to refute an attribution of specified complexity and might even lead to assessments of increasingly extreme complexity. What's more, there's an underlying fact of the matter about what probabilities/complexities are inherent in nature, and this fact of the matter might just be that the complexity of a biotic system is indeed as extreme as it now appears. Why, then, does Van Till think it is "less likely" that specified complexity will be borne out for biotic systems "the more we learn"? The likelihood to which Van Till is referring here has nothing to do with objective assignments of probability or complexity to biotic systems. Rather, this likelihood merely expresses Van Till's personal conviction that naturalistic explanations must inevitably triumph. Any such likelihood is thus purely subjective and flows from Van Till's precommitment to naturalism.

Far from vindicating naturalism, the progress of science, when considered historically and objectively, flings the door wide open to intelligent design. Indeed, the evidence for intelligent design in biology is destined to grow ever stronger. Intelligent design works from the bottom up, finding instances of biological design on a system-by-system basis. Intelligent design does not need to show that every aspect of biology is designed. It is enough to find some clear instances of design and nail them down. The case for intelligent design is therefore cumulative and multipronged. Naturalism, by contrast, must work from the top down, denying that any appearance of design is actual. Naturalism has a vast front to protect. Allow even one exception where design proves actual rather than merely apparent, and the whole edifice of naturalism will come tumbling down. That's why naturalism needs to define design out of existence and cannot afford to weigh the actual biological evidence for design. That evidence is pointing increasingly to the reality of design in nature.

PART 5

THEORETICAL
CHALLENGES TO
INTELLIGENT
DESIGN

30

ARGUMENT FROM IGNORANCE

In attributing design to biological systems, isn't intelligent design just arguing from ignorance?

THIS IS PERHAPS THE MOST COMMON OBJECTION against intelligent design, and it goes by many names: argument from ignorance, argument from silence, argument from personal incredulity, god-of-the-gaps, negative argumentation, argument by elimination, eliminative induction, failure to provide a positive alternative and so on. The underlying concern here is that design theorists argue for the truth of design simply because design has not been shown to be false. In arguments from ignorance, the lack of evidence against a proposition is used to argue for its truth. Here's a stereotypical argument from ignorance: "Ghosts and goblins exist because you haven't shown me that they don't exist."

The argument-from-ignorance objection has been spectacularly successful at shutting down discussion about intelligent design. In fact, among Western intellectuals it functions like a mantra. One merely repeats it whenever the question of design is raised. Design is thus put to flight and calm is restored. Here's a typical exchange between two such intellectuals:

"What do you think about those crazy design theorists?"

"Oh, I've looked at their writings. They're just committing a classic argument from ignorance."

"Yeah, just the same old god-of-the-gaps."

"Hume taught us all about that and thoroughly refuted design." (Conjuring with the name of David Hume is always useful in these discussions.)

"You got it. Hey, let's grab a beer."

"Great idea!"

And thus design is refuted. Perhaps that's why Australian philosopher Alan Olding, in commenting on the persistent use of the argument-from-ignorance or god-of-the-gaps objection against the work of Michael Denton and Michael Behe, writes, "The phrase 'god of the gaps' is nothing more than a question-begging insult meant to stop the flow of argument before it has barely started." (See his article "Maker of Heaven and Micro-biology," *Quadrant*, January-February 2000.)

To see that the argument-from-ignorance objection is not a magic wand for silencing intelligent design, let's begin with a reality check. When the argument-from-ignorance objection is raised against intelligent design, who exactly is accused of being ignorant? It's natural to think that the ig-norance here is on the part of design theorists, who want to attribute intel-ligent agency to biological systems. If only those poor design theorists un-derstood biology better, those systems would readily submit to mechanistic explanation. Thus, when I lecture on university campuses about intelligent design, a biologist in the audience will often get up dur-ing the question-and-answer time to inform me that just because I don't know how complex biological systems might have formed by the Darwin-ian mechanism doesn't mean it didn't happen that way. I then point out that the problem isn't that I personally don't know how such systems might have formed but that the biologist who raised the objection doesn't know how such systems might have formed—and that despite having a fabulous education in biology, a well-funded research laboratory, decades to put it all to use, security and prestige in the form of a tenured academic appointment, and the full backing of the biological community, which has also been desperately but unsuccessfully trying to discover how such sys-tems are formed for more than one hundred years.

Who is ignorant here? Not just the design theorists, but the scientific community as a whole. In fact, it's safe to say that the biological commu-nity is clueless about the emergence of biological complexity. How so? Because the material mechanisms to which the biological community looks to explain biological complexity provide no clue for how those sys-tems might realistically have come about. The problem, therefore, is not ignorance or personal incredulity but *global disciplinary failure* (the disci-pline here being biology) and *gross theoretical inadequacy* (the theory here being Darwin's).

Now, such vast ignorance is not something one typically wants to ad-vertise. A few biologists, however, have now come clean. These include James Shapiro and Franklin Harold, neither of whom supports intelligent design. In a review of Michael Behe's book *Darwin's Black Box* (*National Re-view*, September 16, 1996), James Shapiro, a molecular biologist at the Uni-versity of Chicago, conceded that there are no detailed Darwinian accounts for the evolution of any fundamental biochemical or cellular system, only a variety of wishful speculations. It is remarkable that Darwinism is accepted as a satisfactory explanation for such a vast subject—evolution—with so little rigorous examination of how well its basic theses work in illuminating specific instances of biological adaptation or diversity.

Five years later, cell biologist Franklin Harold wrote a book for Oxford University Press titled *The Way of the Cell*. In virtually identical language, he noted, "There are presently no detailed Darwinian accounts of the evolution of any biochemical or cellular system, only a variety of wishful speculations."

David Ray Griffin, also no supporter of intelligent design, is a philosopher of religion with an interest in biological origins. Commenting on the evolutionary literature that purports to explain how evolutionary transitions lead to increased biological complexity, he writes (in his book *Religion and Scientific Naturalism*),

There are, I am assured, evolutionists who have described how the transitions in question could have occurred. When I ask in which books I can find these discussions, however, I either get no answer or else some titles that, upon examination, do not in fact contain the promised accounts. That such accounts exist seems to be something that is widely known, but I have yet to encounter someone who knows where they exist.

At a recent debate with Brown University biologist Kenneth Miller, I quoted Franklin Harold on the absence of detailed Darwinian accounts for the evolution of complex biological systems. Miller did not challenge Harold's claim. Instead, he impugned Harold's credibility by remarking that Harold was old, having retired fifteen years ago. Presumably Miller was implying that Harold's age put him out of touch with current biological research. But if so, wouldn't the editors at Oxford University Press have declined to publish Harold's book? Age used to command authority and respect, and it still does, so long as one remains faithful to Darwinian orthodoxy. (Witness the homage paid to Ernst Mayr, who is ancient.) But if you challenge biology's most sacred icon—Darwinism—or even state unpleasant facts about it as a loyalist merely wishing to be honest about the situation, don't expect respectful treatment, whatever your age or accomplishments.

Of course the issue is not Harold's age but the substance of his claim. Is he right? Is it the case that there are presently no detailed Darwinian accounts for the evolution of any biochemical or cellular systems but only a variety of wishful speculations? If Harold was wrong, if he had missed the cogent, detailed accounts during the last fifteen years since his retirement, then why didn't Miller point me toward those accounts? Better yet, why doesn't Miller reproduce such detailed accounts in his own writings? To be sure, in his book *Finding Darwin's God* Miller cites what he calls "glittering examples" of Darwinian evolution producing biological complexity. But on closer examination, Griffin's remarks apply all too well to Miller's "glittering examples." Go to the actual papers and books that Miller cites, and the promised accounts of how biological complexity could emerge by Darwinian means are simply not there. (I detail this in chapter five of my book *No Free Lunch*.)

Larry Moran, a molecular biologist at the University of Toronto, is as disingenuous as Miller in claiming that evolutionary biology has resolved the problem of biological complexity. For instance, he asserts that there are "lots and lots of ideas about how irreducibly complex systems arose by evolution" (talk.origins newsgroup, May 2002). Yet when I challenged him to list them, he refused to elaborate. Actually, Moran is probably right that there are "lots and lots of ideas" about how evolution might produce biological complexity. The problem is that invariably these ideas are wishful speculations and not detailed proposals that can be tested. Evolutionary biology is now a field where imagination runs riot and substitutes for rigor.

The biological system that has received the most press from the intelligent design movement is the bacterial flagellum. The bacterial flagellum is a bidirectional motor-driven propeller that sits on the backs of certain bacteria to propel them through their watery environments. It is a marvel of nano-engineering. Howard Berg at Harvard calls it the most efficient machine in the universe. So how did it come about? Let's do a little speculating. It turns out that ten or so genes from the flagellum are homologous to the genes that code for a certain type of pump (known as a type III secretory system). Was this pump therefore an evolutionary precursor to the flagellum? Perhaps, though Milton Saier at the University of California at San Diego argues on the basis of phylogenetic sequence comparisons that the type III secretory system evolved from the flagellum and not vice versa. (And since the flagellum has considerably more complexity than the type III secretory system, explaining the latter in terms of the former is no help in explaining the emergence of biological complexity.) My point here is that just because one can find functional subsystems within some bigger system doesn't mean that the system evolved from these subsystems by Darwinian or other naturalistic means. Similarly, one can't argue that a motorcycle evolved by Darwinian means simply by pointing to a motor and a bicycle. In either case one would need a detailed, testable Darwinian pathway before a Darwinian account of their emergence could be justified.

But hasn't the biological community explained the evolution of such complicated structures as the mammalian eye? Actually, it hasn't. What the biological community has done is noted that there are many different eyes exhibiting varying degrees of complexity—everything from the full mammalian eye at the high end of the complexity scale to a mere light-sensitive spot at the low end. But slapping down eyes of varying complexity on a chart and then drawing arrows from less complex to more complex eyes to signify evolutionary relationships does nothing to explain how increasingly complex eyes emerged. The gaps between these increasingly complex eyes become unbridgeable chasms once you begin to think like an engineer and actually look at the astonishing and irreducibly complex components. To be sure, one can spin a Darwinian tale about how eyes of increasing complexity conferred an advantage in fitness on the organisms that possess them and thus led to the evolution of the mammalian eye. But there's nothing here that you can take to an engineer and use to build an actual eye. Explanations by definition are supposed to clarify and elucidate, engender understanding, and yield practical know-how. Darwinian explanations, like those for the eye, do nothing like this. They are just-so stories—fictional tales that entertain and lull the Darwinian faithful into thinking they've resolved the problem of biological complexity when in fact its solution continues to elude them.

Thus, whenever I hear Darwinists claim that intelligent design constitutes an argument from ignorance, I'm reminded of the 1960s movie *A Guide for the Married Man,* starring Robert Morse and Walter Matthau. Morse takes Matthau under his wing to show him the finer art of infidelity. When Matthau asks Morse what to do if his wife catches him with another woman, Morse says, "Deny, deny, deny." Morse then relates the story of a man whose wife found him in bed with another woman. The scene pans and we see Joey Bishop in bed with a woman and his wife barging into the bedroom and standing over them. When his wife demands to know who the woman is and what she is doing in their bed, Bishop simply keeps asking "What?" (as in "What's the problem?"). Eventually the other woman leaves. The husband (now dressed) ensconces himself in a comfortable chair and reads a newspaper. Confused, his wife changes the subject and asks what he wants for dinner.

Darwinists, in leveling the argument-from-ignorance objection, go this one better. Not only do they deny that there is any problem with their theory, but they turn the tables on anyone who disagrees, attributing fault to Darwin's critics rather than to Darwin's theory. If the Bishop character from the above film had behaved this way, he would have not only feigned ignorance of the other woman in bed with him, but also turned on his wife and started accusing her of adultery. In Freudian terms, Darwinists who accuse intelligent design of arguing from ignorance are guilty of projection as well as denial. Darwinists haven't a clue how systems like the bacterial flagellum might have evolved. On the other hand, we know that intelligence is capable of designing high-tech systems like this. Yet it is the design theorists who are guilty of arguing from ignorance and the Darwinists who know what really happened. The irony here is delicious.

There is much more to intelligent design than an argument from ignorance. But even if all intelligent design had going for it were an argument from ignorance, that would itself be significant. Darwin's theory is widely purported to have resolved the problem of biological complexity. We now find that this main claim to fame is unsupported. Nor are there any other material mechanisms waiting in the wings that promise to pick up where the Darwinian mechanism leaves off. Committed materialists will no doubt think that intelligent design is overemphasizing biology's current problems and that a materialistic solution can be found in time. Yet the fact remains that there are no detailed, testable models for how known material mechanisms can generate biological complexity—only a variety of wishful speculations.

31

ELIMINATIVE INDUCTION

*If the design inference isn't just an
argument from ignorance, how is it more
than an argument from ignorance?*

IF THE ARGUMENT-FROM-IGNORANCE OBJECTION has any force against intelligent design, it doesn't reside in the biological community's ignorance over how material mechanisms gave rise to biological complexity. That ignorance is complete and hardly recommends evolutionary biology as a field offering profound insight into the emergence of biological complexity. Rather, the argument-from-ignorance objection gains force from the faulty logic that is supposed to underlie it. Critics charge that intelligent design is based on a purely negative form of argumentation. Accordingly, everything depends on establishing that the origin of certain biological systems defies naturalistic explanation. Once this negation is in hand, intelligent design is said to flip-flop, illegitimately transforming this negation into the affirmation that these systems therefore had to be designed. Thus design theorists are supposed to be guilty of reasoning directly from the premise "Shucks, no one has figured out how the flagellum arose" to the conclusion "Gee, it must have been designed."

Kenneth Miller, for instance, makes this charge in his article "The Flagellum Unspun" (see <www.millerandlevine.com/km/evol/design2/article.html>). Miller, despite a long exposure to intelligent design thinkers and writings, continually misses a crucial connecting link in the argument. Let me therefore spell out the premises of the argument as well as its conclusion: Certain biological systems have a feature, call it SC (specified complexity). Darwinians don't have a clue how biological systems with that feature originated. (Miller disputes this premise but, as we saw in the last chapter, without effect.) We know that intelligent agency has the causal power to produce systems that exhibit SC (e.g., many human artifacts exhibit SC)—*this is the crucial connecting premise.* Therefore, biological systems that exhibit SC are likely to be designed. Design theorists, in attributing design to systems that exhibit SC, are simply doing what scientists do generally, which is attempt to formulate a causally adequate explanation for the phenomenon in question.

To attribute specified complexity, and thereby design, to a biological system is to engage in an eliminative induction, a form of reasoning used throughout the sciences. Eliminative inductions argue for the truth of a proposition by arguing that competitors to that proposition are false. (Contrast this with Popperian falsification, where propositions are corroborated to the degree that they successfully withstand attempts to falsify them.) Provided the proposition, together with its competitors, form a mutually exclusive and exhaustive class, eliminating all the competitors entails that the proposition is true. (Recall Sherlock Holmes's famous dictum: "When you have eliminated the impossible, whatever remains, however improbable, must be the truth.") This is the ideal case, in which eliminative inductions in fact become deductions. The problem is that in practice we don't have a neat ordering of competitors that can then all be knocked down with a few straightforward and judicious blows (like pins in a bowling alley). In *Bayes or Bust*, philosopher of science John Earman puts it this way:

> The eliminative inductivist [seems to be] in a position analogous to that of Zeno's archer whose arrow can never reach the target, for faced with an infinite number of hypotheses, he can eliminate one, then two, then three, etc., but no matter how long he labors, he will never get down to just one. Indeed, it is as if the arrow never gets half way, or a quarter way, etc. to the target, since however long the eliminativist labors, he will always be faced with an infinite list [of remaining hypotheses to eliminate].

Earman offers these remarks in a chapter titled "A Plea for Eliminative Induction." He himself thinks there is a legitimate and necessary place for eliminative induction in scientific practice. What, then, does he make of this criticism? Here is how he handles it:

> My response on behalf of the eliminativist has two parts. (1) Elimination need not proceed in such a plodding fashion, for the alternatives may be so ordered that an infinite number can be eliminated in one blow. (2) Even if we never get down to a single hypothesis, progress occurs if we succeed in eliminating finite or infinite chunks of the possibility space. This presupposes, of course, that we have some kind of measure, or at least topology, on the space of possibilities.

To this Earman adds that eliminative inductions are typically *local inductions*. For such inductions there is no pretense of considering all logically possible hypotheses. Rather, there is tacit agreement on the explanatory domain of the hypotheses as well as on what auxiliary hypotheses may be used in constructing explanations.

That's why intelligent agency having the causal power to produce systems that exhibit specified complexity is such an important premise in eliminative inductions that attempt to infer biological design. Let's even give this premise a name: *the can-do premise* (because we know that designers "can do" it, that is, they can generate specified complexity). Precisely because intelligent agency is reliably correlated with specified complexity, there is no need to give equal weight to every conceivable naturalistic hypothesis or to wade interminably through the never-ending list of half-baked, handwaving Darwinian just-so stories, none of which has ever given any evidence of actually elucidating biological systems that exhibit specified complexity. In other words, the can-do premise makes the eliminative induction here a local induction that can legitimately infer design.

Design critics need also to take to heart Earman's claim that eliminative inductions can be *progressive*. Too often specified complexity is charged with underwriting a purely negative form of argumentation. But that charge is not accurate. The argument for the specified complexity of the bacterial flagellum, for instance, makes a positive contribution to our understanding of the limitations that natural mechanisms face in trying to account for it. Eliminative inductions, like all inductions and indeed all scientific claims, are fallible. But they need a place in science. To refuse them, as evolutionary biology tacitly does by rejecting specified complexity as a criterion for detecting design, does not keep science safe from disreputable influences but instead undermines scientific inquiry itself.

The way things stand now, evolutionary biology has set in place procedural rules that allow intelligent design only to fail but not to succeed. If evolutionary biologists can discover or construct detailed, testable, indirect Darwinian pathways that account for complex biological systems like the bacterial flagellum, then intelligent design will rightly fail. On the other hand, evolutionary biology makes it effectively impossible for intelligent design to succeed. According to evolutionary biology, intelligent design has only one way to succeed, namely, by showing that complex specified biological structures could not have evolved via any material mechanism. In other words, so long as some unknown material mechanism might have evolved the structure in question, intelligent design is proscribed.

Evolutionary theory is thereby rendered immune to disconfirmation in principle because the universe of unknown material mechanisms can never be exhausted. Indeed, the evolutionist has no burden of evidence. But notice that there's no logically consistent reason evolutionists shouldn't hold themselves to the same ridiculous burden of evidence. Indeed, if any side should have to shoulder that impractical burden, history tells it should be the evolutionary naturalists. Why? They are in the extreme, historical minority in denying that biological systems are designed. More significantly, they themselves admit that biological systems appear on the face of it to belong to that known class of things that are intelligently designed. For instance, in *What Mad Pursuit* Francis Crick writes, "Biologists must constantly keep in mind that what they see was not designed, but rather evolved."

If a creature looks like a dog, smells like a dog, barks like a dog, feels like a dog and pants like a dog, the burden of evidence lies with the person who insists the creature isn't a dog. The same goes for incredibly intricate machines like the bacterial flagellum: the burden of evidence is on those who want to deny its design. And yet you won't find Darwinists rolling up their sleeves and trying to eliminate every imaginable and as yet unimagined intelligent design scenario, pleading for patience while they work their way through an infinite set of possibilities. Instead, they spuriously shift the burden of evidence entirely to the skeptic of naturalistic evolution, insisting that these skeptics establish a universal negative not merely by an eliminative induction (such inductions are invariably local and constrained) but by an exhaustive search and elimination of all conceivable naturalistic possibilities—however remote, however unfounded, however unsupported by evidence. That is not how science is supposed to work.

Science is supposed to give the full range of possible explanations a fair chance to succeed. That's not to say that anything goes; but it is to say that anything might go. In particular, science may not by a priori fiat rule out logical possibilities. Evolutionary biology, by limiting itself exclusively to material mechanisms, has settled in advance which biological explanations are true apart from any consideration of empirical evidence. This is armchair philosophy. Intelligent design may not be correct. But the only way we could discover that is by admitting design as a real possibility, not by ruling it out a priori. Darwin himself would have agreed. In the *Origin of Species* he wrote, "A fair result can be obtained only by fully stating and balancing the facts and arguments on both sides of each question."

32

HUME, REID AND
SIGNS OF INTELLIGENCE

*Didn't David Hume demolish not just the
design argument for the existence of God
but also any sort of inference to design
based on features of the natural world?*

DAVID HUME'S CRITIQUE OF INTELLIGENT DESIGN is vastly overrated.
Nevertheless, his critique, especially in the hands of his contemporary disciples, has been highly effective at shutting down discussion about design.
Here I want to review Hume's critique, indicate how modern disciples
have updated it and then describe the response to Hume by his contemporary Thomas Reid. Reid's response, in my view, is decisive. Would that
more philosophers studied it. Hume did not demolish design. Reid demolished Hume.

Hume's critique of design is found in his *Dialogues Concerning Natural
Religion*, published in 1779, three years after his death. Hume's case
against design, unlike Darwin's, is purely philosophical. Darwin argues
against design on scientific grounds by claiming to provide a natural
mechanism that could account for the appearance of design in nature.
Hume, on the other hand, argues against design by claiming to find logical
flaws with it. Hume rightly shows that British natural theologians were
overselling the design argument. Indeed, there is no valid inferential chain
from the appearance of design in nature to the main character of the Bible
or even to some stripped down deistic version of this biblical God. Hume
argues that even if a designer could be inferred from the appearance of design in nature, such a designer's goodness, wisdom and plurality (i.e.,
whether there be one or many) could not.

Design inferences based on the appearances of design in nature need to
be modest. Hume was not alone in urging such modesty. Immanuel Kant

argues that at best the design argument could infer a designer responsible for designs within nature but not a creator God responsible for nature as such (see chapter seven). Even Thomas Aquinas admits the need for modesty in design reasoning. In his *Summa Contra Gentiles*, Aquinas wrote,

> By his natural reason man is able to arrive at some knowledge of God. For seeing that natural things run their course according to a fixed order, and since there cannot be order without a cause of order, men, for the most part, perceive that there is one who orders the things that we see. But who or of what kind this cause of order may be, or whether there be but one, cannot be gathered from this general consideration.

Aquinas here is not doing first philosophy or metaphysics. He is simply noting that our natural reason readily infers some sort of "orderer" or "designer" behind nature. Aquinas calls this designer "God," but he is clearly speaking of this designer very loosely— for Aquinas, the nature and even plurality of that designer could not be settled simply by studying nature.

In his *Dialogues Concerning Natural Religion,* however, Hume goes beyond urging modesty in design arguments. He attacks even a modest inference to an unspecified designer. His first main criticism is that, at best, design constitutes a weak argument from analogy. His other major criticism is that design fails as an inductive generalization. Both these criticisms miss the mark. Consider first Hume's criticism of design as a weak argument from analogy. The problem with arguments from analogy is that they are always also arguments from disanalogy. Indeed, if there were no disanalogy, there would be no need to argue from analogy, because in that case we would be dealing with things that are identical and not merely analogous. (Things analogous in every respect are identical.)

Arguments from analogy argue that two things share some feature because they share some other features (which constitute the basis of the analogy). For instance, consider a watch and an organism. We know the watch is designed. We also know that watches and organisms share certain features (like functional interdependence of parts, adaptation of means to ends, self-propulsion, etc.). Given these shared features, is it fair to conclude that organisms are designed? The problem is that watches and organisms also diverge on some features. Watches are made of metal and glass; organisms are not. Organisms repair themselves; watches do not. The million-dollar question, therefore, is whether design is a shared feature of watches and organisms (like functional interdependence of parts) or a divergent feature (like self-repair). According to Hume, there's no way to decide simply on the basis of such analogical and disanalogical information.

Even so, there is a way to strengthen the argument from analogy, and that is by arguing that the features shared by the items in question have never in our experience been divorced from the feature in question. Suppose that the items in question are watches and organisms and that the feature in question is design. If it could be shown that features shared by watches and organisms—like functional interdependence of parts, adaptation of means to ends and self-propulsion—have in our experience always resulted from the work of a designing intelligence, then it would be reasonable, as an inductive generalization, to conclude that organisms, like watches, are designed. Schematically the argument would look as follows. (P_1, P_2 and P_3 are the premises; C is the conclusion.)

P_1: Watches are designed.

P_2: Watches and organisms exhibit functional interdependence of parts, adaptation of means to ends, and so on.

P_3: There is no known instance where something exhibits functional interdependence of parts, adaptation of means to ends, and so on, without being designed.

C: Therefore, organisms are designed as well.

Although reframing the design argument as an inductive generalization turns it into a valid argument from analogy, this reframing runs smack into Hume's second objection. Hume and the Humean tradition reject such inductive generalizations. The problem is that inductive generalizations are supposed to be based on past experience. And while we have past experience of watches being designed, Hume would claim that we have no experience of organisms, or for that matter a universe, being designed. Hume's modern disciples agree. Robert Pennock, for instance, remarks that design inferences must be "based upon known types of causal processes" ("The Wizards of ID," *Intelligent Design Creationism*, MIT Press, 2001). He therefore claims that design inferences become more tenuous as the underlying causal processes become less well known.

> When archeologists pick out something as an artifact or suggest possible purposes for some unfamiliar object they have excavated, they can do so because they already have some knowledge of the causal processes involved and have some sense of the range of purposes

that could be relevant. It gets more difficult to work with the concept when speaking of extraterrestrial intelligence, and harder still when considering the possibility of animal or machine intelligence. But once one tries to move from natural to supernatural agents and powers as creationists desire, "design" loses any connection to reality as we know it or can know it scientifically.

And for Pennock, as for fellow Humeans generally, if it can't be known scientifically, then it can't be known at all. (Hume, after all, consigned metaphysics to the flames.)

Wesley Elsberry and John Wilkins make essentially the same point. They maintain that there are "two kinds of design—the ordinary kind based on a knowledge of the behavior of designers, and a 'rarefied' design, based on an inference from ignorance, both of the possible causes of regularities and of the nature of the designer" (see "The Advantages of Theft over Toil: The Design Inference and Arguing from Ignorance," in *Biology and Philosophy*, vol. 16, 2001). Accordingly, a design inference that infers design merely by looking at certain features of an object without knowing anything about its underlying causal story cannot infer ordinary design but only rarefied design. For Elsberry and Wilkins rarefied design means an attribution of design based on an absence of naturalistic alternatives and thus serving merely as a stop-gap for ignorance.

Such objections by Pennock, Elsberry and Wilkins are typical of the Humean inductive tradition. Accordingly, to know that an object is designed we first need to know something about the designer. Since the Humean tradition is committed to empiricism, the first thing we need is direct observational experience of the designer or one like it (which implies the designer needs to be physically embodied). We also need to know something about the capacities of the designer to bring about design. And finally we need to know something about the designer's purposes and motives, for how else could we predict whether a designer would be likely to bring about a given design? As Elsberry and Wilkins put it, design within the Humean inductive tradition is a "form of causal regularity that may be adduced to explain the probability of an effect being high, and which depends on a set of background theories and knowledge claims about designers."

All of these restrictions on inferring design are, of course, very convenient for keeping designers unacceptable to naturalism at bay. Indeed, there's no way for a transcendent designer to get a foot in the door once one accepts this Humean inductive framework for design reasoning. But why should one accept this framework in the first place? It seems on its face an exercise in special pleading. Consider the search for extraterrestrial intelligence (SETI). If we were to receive a radio signal from outer space representing a long sequence of prime numbers (as in the movie *Contact*), we would know we were dealing with an intelligence—indeed, SETI researchers would be dancing in the streets, the *New York Times* would be trumpeting the discovery, and Nobel Prizes would duly be awarded.

But what exactly would we know about the intelligence responsible for that signal? Suppose all we had was this signal representing a sequence of primes. Would we know anything about the intelligence's purposes and motives for sending the primes? Would we know anything about the technology it employed? Would we know anything about its physical makeup? Would we even know that it was physical? Our evidence for design in this case would be entirely circumstantial. We would be confronted with an effect but be unable to trace back to its cause.

Consider a more extreme example still. Imagine a device that outputs zeroes and ones for which our best science tells us that the bits are independent and identically distributed with uniform probability. (The device is therefore an idealized coin-tossing machine. Note that quantum mechanics offers such a device in the form of photons shot at a polarizing filter whose angle of polarization is 45 degrees in relation to the polarization of the photons: half the photons will go through the filter, counting as a "one"; the others will not, counting as a "zero.") Now, what happens if we control for all possible physical interference with this device, and nevertheless the bit string that this device outputs yields an English text-file in ASCII code that resolves outstanding mathematical problems, explains the cure for cancer and delineates previously unimagined technologies? The output of this device is therefore not only designed (and obviously so) but also exceeds all current human design. Yet our best science has no way of prescribing a causal account for how this design was imparted. By Hume's logic, we would have to shrug our shoulders and say, "Golly, isn't nature amazing!"

The fact is that we infer design repeatedly and reliably without knowing characteristics of the designer or being able to assess what a designer is likely to do. Humeans in their weaker moments admit as much. Take Elliott Sober. Before he permits intelligent design into biology, he wants to know the characteristics of the designer, the independent evidence for the existence of that designer and what sorts of biological systems we should expect from such a designer. According to Sober, if the design theorist cannot answer these questions, then intelligent design is untestable and therefore unfruitful for science. Yet in a footnote that deserves to be part of his main text, Sober admits,

> To infer watchmaker from watch, you needn't know exactly what the watchmaker had in mind; indeed, you don't even have to know that the watch is a device for measuring time. Archaeologists sometimes unearth tools of unknown function, but still reasonably draw the inference that these things are, in fact, *tools*. ("Testability," 1999 presidential address to the American Philosophical Association)

Because he is wedded to the Humean inductive tradition, Sober views all our knowledge of the world as an extrapolation from past experience. Thus for design to be explanatory, it must fit our preconceptions; and if it does not, it must lack empirical justification. For Sober, to predict what a designer would do requires first looking to past experience and determining what designers in the past have actually done. And yet his comment about watchmakers and watches belies such a view, for he admits we could know that watches were designed even if we knew nothing about watchmakers and that mysterious tools were designed even if we knew nothing about the toolmakers or even the precise function of the tools. Within the Humean inductive tradition, designers are in the same boat as natural laws, with their explanatory power located in an extrapolation from past experience. To be sure, designers, like natural laws, can behave predictably. (Designers often institute *policies* that other designers then dutifully obey.) Yet unlike natural laws, which are universal and uniform, designers are also innovators. Innovation, the emergence of true novelty, eschews predictability. It therefore follows that design cannot be subsumed within a Humean inductive framework. Designers are inventors. We cannot predict what an inventor would do short of becoming that inventor.

But the problem goes deeper. Not only can't Humean induction tame the unpredictability inherent in design, but it can't account for how we recognize design in the first place. Sober, for instance, regards the design hypothesis for biology as fruitless and untestable because it fails to confer an ascertainable probability on biologically interesting propositions. But take a different example, say from archeology, in which a design hypothesis about certain aborigines predicts certain artifacts, say arrowheads. Such a design hypothesis would, on Sober's account, be testable and thus acceptable to science. But what sort of archeological background knowledge had to go into that design hypothesis to make it a successful predictor of arrowheads? At the very least, we would need past experience with arrowheads. But how did we recognize that the arrowheads in our past experience were designed? Did we see humans actually manufacture those arrowheads? If so, how did we recognize that these humans were acting deliberately as designing agents and were not just randomly chipping away at random chunks of rock? (Carpentry and sculpting entail design; but whittling and chipping, though performed by intelligent agents, do not.) As is evident from this line of reasoning, the induction needed to recognize design can never get started. Our ability to recognize design must therefore arise independently of induction and, thus, independently of a Humean inductive framework.

That was precisely Reid's point, and in making it he demolished once and for all Humean induction as applied to design. In 1780, only a year following the publication of Hume's *Dialogues Concerning Natural Selection*, Reid delivered a set of lectures on natural theology in Glasgow (reprinted in *Lectures on Natural Theology*, University Press of America, 1981). In those lectures he remarked,

> No man ever saw wisdom [read "design" or "intelligence"], and if he does not [infer wisdom] from the marks of it, he can form no conclusions respecting anything of his fellow creature. How should I know that any of this audience have understanding? It is only by the effects of it on their conduct and behavior, and this leads me to suppose that such behavior proceeds only from understanding. But says Hume, unless you know it by experience, you know nothing of it. If this is the case, I never could know it at all. Hence it appears that whoever maintains that there is no force in the argument from final causes [design], denies the existence of any intelligent being but himself. He has the same evidence for wisdom and intelligence in God as in a father or brother or a friend. He infers it in both from its effects and these effects he discovers in the one as well as the other. . . . From marks of wisdom and intelligence in effects, a wise and intelligent cause may be inferred.

According to Reid, we attribute design as an inference from signs of intelligence (or "from marks of intelligence and wisdom in effects," as he put it). We do not get into the mind of designers and thereby attribute design. Rather, we recognize their intelligence by examining the effects of their actions and determining whether those effects display signs of intelligence. Accordingly, when one purports to attribute design on the basis of induction, one has already presupposed the ability to identify design independently of induction.

Take an anthropologist watching a native islander chipping stones. Is the native an arrowhead maker and therefore a designer? If our anthropologist saw the native banging away at a stone with a second rock ideal for chipping arrowheads, all the while gazing most seriously at the stone, this by itself would not prove that the native was designing something. Even if the native, upon seeing the anthropologist, began lecturing in exquisite and engrossing English on the ancient art of arrowhead making, the anthropologist would still not know that the stone getting beaten was a designed object, much less an arrowhead. If the anthropologist looked down during the exquisite lecture and found that the stone had been beaten to dust that was then carried off by the wind, for all the anthropologist could gather from this seemingly useless dust, the native might have been banging stones merely to relieve frustration. If, on the other hand, the anthropologist looked down at the end of the native's exquisite lecture and found an exquisite arrowhead fit for a king, this, rather than the native's motions or words, would demonstrate that the object produced was in fact a designed object and that the native was in fact an arrowhead maker and therefore a designer.

In short, we recognize intelligence by its effects, not by directly perceiving it. A human being who continuously mumbles the same nonsense syllable displays no intelligence and provides no justification for attributing design. Design reasoning is effect-to-cause reasoning: it begins with effects in the physical world that exhibit clear signs of intelligence and from those signs infers to an intelligent cause. Neither of Hume's two main criticisms against design therefore holds up. Induction is entirely the wrong analytic framework for how we infer design. And Hume's concern about design inferences involving faulty analogies is misconceived. The signs of intelligence that occur in human artifacts and biological systems are not merely analogous. They are isomorphic, for we find the exact same form of specified complexity in each.

The very idea that there could be something like a sign of intelligence (much less that it could be given analytic precision via specified complexity) is anathema to the Humean inductive tradition. Signs of intelligence, by their very nature, do not submit to Humean induction. Yet as Reid showed, though signs of intelligence can be learned and confirmed by experience, our ability to recognize them cannot originate in experience. That ability is hardwired into us as part of basic human rationality. It is, as Alvin Plantinga would put it, part of our "proper function." Hume and his followers exercise that proper function and do so daily, just like everyone else. What's new with the contemporary intelligent design movement is that it brings analytic precision to our understanding of these signs of intelligence. Within the theory of intelligent design, signs of intelligence get cashed out as specified complexity, which serves as an analytic tool for scientifically assessing whether design actually is present in various phenomena.

33

DESIGN BY ELIMINATION VERSUS DESIGN BY COMPARISON

How are design hypotheses properly inferred—simply by eliminating chance hypotheses or by comparing the likelihood of chance and design hypotheses?

BEHIND THIS QUESTION ARE TWO FUNDAMENTALLY different approaches about how to reason with chance hypotheses, one friendly to intelligent design, the other less so. The friendly approach, due to Ronald Fisher, rejects a chance hypothesis provided sample data appear in a prespecified rejection region. The less friendly approach, due to Thomas Bayes, rejects a chance hypothesis provided an alternative hypothesis confers a bigger probability on the data in question than the original hypothesis. In the Fisherian approach, chance hypotheses are rejected in isolation for rendering data too improbable. In the Bayesian approach, chance hypotheses are eliminated provided some other hypotheses render the data more probable. Whereas in the Fisherian approach the emphasis is on elimination, in the Bayesian approach the emphasis is on comparison. These approaches are incompatible, and the statistical community itself is deeply riven over which of these approaches to adopt as the right canon for statistical rationality. The difference reflects a deep divergence in fundamental intuitions about the nature of statistical rationality and in particular about what counts as statistical evidence.

The most influential criticism of specified complexity charges it with falling on the wrong side of this divide. Specifically, critics charge that to use specified complexity to infer design presupposes an eliminative, Fisherian approach to reasoning with chance hypotheses whereas the right approach to inferring design needs to embrace a comparative, Bayesian approach. The most prominent scholar to make this criticism is Elliott Sober.

Other scholars have offered this criticism as well, and many more still have cited it as decisively refuting specified complexity as a sign of intelligence.

In responding to this criticism, let's begin with a reality check. Often when the Bayesian literature tries to justify Bayesian methods against Fisherian methods, authors are quick to note that Fisherian methods dominate the scientific world. For instance, Richard Royall (who, strictly speaking, is a likelihood theorist rather than a Bayesian, though the distinction is not crucial to this discussion) writes, "Statistical hypothesis tests, as they are most commonly used in analyzing and reporting the results of scientific studies, do not proceed . . . with a choice between two [or more] specified hypotheses being made . . . [but follow] a more common procedure" (*Statistical Evidence: A Likelihood Paradigm,* Chapman & Hall, 1997). Royall then outlines that common procedure, which requires specifying a single chance hypothesis, using a test-statistic to identify a rejection region, checking whether the probability of that rejection region under the chance hypothesis falls below a given significance level, determining whether a sample (the data) falls within that rejection region and, if so, rejecting the chance hypothesis. In other words, the sciences look to Fisher and not Bayes for their statistical methodology. Colin Howson and Peter Urbach, in *Scientific Reasoning: The Bayesian Approach,* likewise admit the underwhelming popularity of Bayesian methods among working scientists.

So are the majority of scientists just being stupid or lazy in adopting a Fisherian approach to statistical reasoning? To answer this question, let's look at two prototypical examples where Fisherian and Bayesian methods are employed. Once these examples are in hand, we can tinker with them to see what can go wrong with both methods. Let's start with an example of Fisherian reasoning. The Fisherian approach eliminates chance hypotheses in isolation, so we need only consider a single chance hypothesis for elimination. Let's take a particularly simple one, namely, the chance hypothesis that characterizes the tossing of a fair coin. To test whether the coin is biased in favor of heads (and thus not fair), one can set a rejection region of ten heads in a row and then flip the coin ten times. In Fisher's approach, if the coin lands ten heads in a row, then one is justified in rejecting the chance hypothesis. The improbability of tossing ten heads in a row, assuming the coin is fair, is approximately one in a thousand (i.e., .001).

Next, to illustrate the Bayesian approach, consider the following probabilistic setup. Imagine two coins, the one fair and the other biased. Assume the biased coin has a probability of landing heads 90 percent of the time. In addition, imagine a giant urn with a million equally sized balls, all of which except one are white, the lone exception being black. Now imagine that a single random sample will be taken from the urn and that if a white ball is selected (which is overwhelmingly probable), then the fair coin will be tossed ten times; but if the one lone black ball is selected (which is overwhelmingly improbable), then the biased coin will be tossed ten times. Now imagine that all you see is a coin tossed ten times and each time landing heads. The probability of it landing ten heads in a row given that the fair coin was tossed is approximately .001 (one in a thousand). But the probability of it landing ten heads in a row given that the biased coin was tossed is approximately .35 (a little better than one in three). Within the Bayesian literature, these probabilities are known as *likelihoods*.

So which coin was tossed, the fair one or the biased one? If one looks purely at likelihoods, it appears that the biased coin was tossed; indeed, it's much more likely that ten heads in a row will appear using the biased coin rather than the fair coin. But that answer will not do. The problem is that which coin gets tossed has what in the Bayesian literature is called a *prior probability.* That prior probability renders it much more likely that the fair coin was tossed than that the biased coin was tossed. The fair coin has prior probability .999999 of being tossed (because a white ball is that likely to be selected from the urn), whereas the biased coin has prior probability .000001 of being tossed (because the one lone black ball is only that likely to be selected from the urn).

To decide which coin was tossed, these prior probabilities need to be factored into the likelihoods calculated earlier. To do that, one calculates what in the Bayesian literature are known as *posterior probabilities.* (These are calculated using Bayes's theorem.) The posterior probability of the fair coin being tossed given that ten heads in a row were observed is .9996, whereas the posterior probability of the biased coin being tossed given that ten heads in a row were observed is .0004. Given the probabilistic setup for the two coins and the urn as described above, it is therefore much more probable that the fair coin was tossed than that the biased coin was tossed. And this is the case even though the observed outcome of ten heads in a row taken by itself is more consistent with the biased coin than with the fair coin.

Given these particularly neat and clean illustrations of the Fisherian and Bayesian approaches, one might wonder what could be the problem with either. Both approaches, as illustrated in these examples, seem eminently reasonable given the questions they are called to answer. Nevertheless, both approaches raise serious conceptual problems when probed more deeply. I want, in the remainder of this chapter, to describe the conceptual problems raised by the Fisherian approach and to indicate how my work on specified complexity helps resolve them. Next I want to describe the conceptual problems raised by the Bayesian approach and to indicate why they render it inadequate as a general model of statistical rationality. In particular, I will show how the Fisherian approach can be made logically coherent and why the Bayesian approach, when it works (which is not too often), must in fact presuppose the Fisherian approach.

So what are the problems with the Fisherian approach, and how does my work on specified complexity help resolve them? Schematically, the Fisherian approach looks like this: A chance hypothesis defined with respect to a reference class of possibilities is given. Also given is a rejection region from that reference class. With the chance hypothesis and rejection region in place, an event is then sampled from the reference class of possibilities. If that event (the sample or data) falls within the rejection region, and if the probability of that rejection region with respect to the chance hypothesis is sufficiently small, then the chance hypothesis is rejected. Intuitively, think of an arrow shot at a large wall displaying a fixed target. The wall corresponds to the reference class of possibilities (all the places the arrow might land), and the target corresponds to the rejection region. Provided that the arrow landing in the target (i.e., the sample falling in the rejection region) has sufficiently small probability, then the chance hypothesis is rejected. In our earlier coin-tossing example, the rejection region was all possible sequences of heads and tails, the rejection region was all sequences beginning with ten heads in a row, the sample was a sequence of ten heads in a row, and the chance hypothesis presupposed a fair coin.

Is there something wrong with this picture? Although this picture has proven quite successful in practice, Fisher, in formulating its theoretical underpinnings, left something to be desired. There are three main worries: First, how does one make precise what it means for a rejection region to have "sufficiently small" probability with respect to a chance hypothesis? Second, how does one characterize rejection regions so that a chance hypothesis doesn't automatically get rejected in case it actually is operating? And third, why should a sample that falls in a rejection region count as evidence against a chance hypothesis?

The first concern is usually stated in terms of setting a "significance level." A significance level prescribes the degree of improbability below which a rejection region eliminates a chance hypothesis once the sample falls within it. Significance levels in the social sciences literature, for instance, usually weigh in at .05 or .01. But where do these numbers come from? They are, in fact, entirely arbitrary. This arbitrariness has dogged the Fisherian approach from the start. Nevertheless, there is a way around it.

Consider again our example of tossing a coin ten times and getting ten heads in a row. The rejection region, which matches this sequence of coin tosses, therefore sets a significance level of .001. If we tossed ten heads in a row, we might therefore regard this as evidence against the coin being fair. But what if we didn't just toss the coin ten times on one occasion but tossed it ten times on multiple occasions? If the coin's behavior was entirely what one would expect from a fair coin most of the time we tossed it, then on those few occasions when we observed ten heads in a row, we would have no reason to suspect that the coin was biased since fair coins, if tossed sufficiently often, will produce any sequence of coin tosses, including ten heads in a row. The strength of the evidence against a chance hypothesis when a sample falls within a rejection region therefore depends on how many samples are taken or might have been taken. These samples constitute what I call *replicational resources*. The more such samples, the greater the replicational resources.

Significance levels therefore need to factor in replicational resources if samples that match these levels are to count as evidence against a chance hypothesis. But that's not enough. In addition to factoring in replicational resources, significance levels also need to factor in what I call *specificational resources*. The rejection region on which we've been focusing specified ten heads in a row. But surely if samples that fall within this rejection region could count as evidence against the coin being fair, then samples that fall within other rejection regions must likewise count as evidence against the coin being fair. For instance, consider the rejection region that specifies ten tails in a row. By symmetry, samples that fall within this rejection region must count as evidence against the coin being fair just as much as samples falling within the rejection region that specifies ten heads in a row.

But if that is the case, then what's to prevent the entire range of possible coin tosses from being swallowed up by rejection regions so that regardless of what sequence of coin tosses is observed, it always ends up falling in some rejection region and therefore counting as evidence against the coin being fair? More generally, what's to prevent any reference class of possibilities from being partitioned into a mutually exclusive and exhaustive collection of rejection regions so that any sample will always fall in one of these rejection regions and therefore count as evidence against any chance hypothesis whatsoever?

The way around this concern is to limit rejection regions to those that can be characterized by low-complexity patterns. (Such a limitation has in fact been implicit when Fisherian methods are employed in practice.) Rejection regions, and specifications more generally, correspond to events and therefore have an associated probability or probabilistic complexity. But rejection regions are also patterns, and as such they have an associated complexity that measures the degree of complication of the patterns, or what I call its *specificational complexity.* Typically this form of complexity corresponds to a Kolmogorov compressibility measure or minimum description length. (The shorter the description, the lower the specificational complexity. See <www.mdl-research.org>.) I summarize these two types of complexity in chapter ten. Note that specificational complexity arises very naturally: it is not an artificial or ad hoc construct designed simply to shore up the Fisherian approach. Rather, it has been implicit right along, enabling Fisher's approach to flourish despite the inadequate theoretical underpinnings that Fisher provided for it.

Replicational and specificational resources together constitute what I call *probabilistic resources.* Probabilistic resources resolve the first two worries (raised above) concerning Fisher's approach to statistical reasoning. Specifically, probabilistic resources enable us to set rationally justified significance levels, and they constrain the number of specifications, thereby preventing chance hypotheses from getting eliminated willy-nilly. Probabilistic resources therefore provide a rational foundation for the Fisherian approach to statistical reasoning. What's more, by estimating the probabilistic resources available in the known physical universe, we can set a significance level that's justified irrespective of the probabilistic resources in any given circumstance. Such a context-independent significance level is thus universally applicable and definitively answers what it means for a significance level to be "sufficiently small" regardless of circumstance. For a conservative estimate of this significance level, known as a *universal probability bound,* see chapter ten. For the details about placing Fisher's approach to statistical reasoning on a firm rational foundation, see chapter two of my book *No Free Lunch.*

That leaves the third worry concerning the Fisherian approach to statistical reasoning: why should a sample that falls in a rejection region (or, more generally, an outcome that matches a specification) count as evidence against a chance hypothesis? Once one allows that the Fisherian approach is logically coherent and that one can eliminate chance hypotheses individually simply by checking whether samples fall within suitable rejection regions (or, more generally, whether outcomes match suitable specifications), then it is a simple matter to extend this reasoning to entire families of chance hypotheses, perform an eliminative induction (see chapter thirty-one), and thereby eliminate all relevant chance hypotheses that might explain a sample. And from there it is but a small step to infer design.

Let's stay with this last point for a moment: How does one go from eliminating chance to inferring design? Indeed, what justifies this move from chance elimination to design inference? We are supposing, for the moment, that the Fisherian approach can legitimately eliminate individual chance hypotheses and thus, by successive elimination, eliminate whole families of chance hypotheses. To eliminate a chance hypothesis, the Fisherian approach determines whether an outcome matches a specification and whether the specification itself describes an event of small probability. (The event here comprises all outcomes that match the specification.) Given that we've successfully characterized all chance hypotheses that exclude design and that we've been able to eliminate them by means of such a specification (the outcome therefore exhibits specified complexity), why should we think that outcome is designed?

In this case the specification itself acts as a logical bridge between chance elimination and design inference. Here's the rationale: if we can spot an independently given pattern (i.e., specification) in some observed outcome and if possible outcomes matching that pattern are, taken jointly, highly improbable (in other words, if the observed outcome exhibits specified complexity), then it's more plausible that some end-directed agent or process produced the outcome by purposefully conforming it to the pattern than that it simply by chance ended up conforming to the pattern. Accordingly, even though specified complexity establishes design by means of an eliminative argument, it is not fair to say that it establishes design by means of a *purely* eliminative argument. The independently given pattern, or specification, contributes positively to our understanding of the design inherent in things that exhibit specified complexity.

To avoid this slippery slope to design, Bayesian theorists deny that the Fisherian approach can legitimately eliminate even one chance hypothesis (much less sweep the field clear of all relevant chance hypotheses as required for a successful design inference). The problem, as they see it, is that samples falling within rejection regions (or, more generally, outcomes matching specifications) cannot serve as evidence against chance hypotheses. Rather, the only way for there to be evidence against a chance hypothesis is for there to be better evidence in favor of some other hypothesis.

I'll analyze the Bayesian approach to statistical evidence momentarily, but first I need to say a word about evidence generally. In *World Without Design*, Michael Rea remarks, "True inquiry is a process in which we try to revise our beliefs on the basis of what we take to be evidence." He continues,

> But this means that, in order to inquire into anything, we must already be disposed to take some things as evidence. In order even to begin inquiry, we must already have various dispositions to trust at least some of our cognitive faculties as sources of evidence and to take certain kinds of experiences and arguments to be evidence. Such dispositions (let's call them *methodological dispositions*) may be reflectively and deliberately acquired.

Accordingly, what counts as evidence (and that includes statistical evidence) is decided not on the basis of evidence but on the basis of dispositions that themselves are not mandated by evidence. Why, for instance, do most mathematicians find proof by contradiction (i.e., *reductio ad absurdum*) as compelling evidence for the truth of a mathematical proposition, but others (the intuitionists) find such proofs inadequate and instead require constructive proofs? Or again, why do Fisherian and Bayesian approaches to statistical evidence remain at loggerheads? In such cases the debate is not merely over how to weigh certain evidence but over what counts as evidence in the first place. The issue of what counts as evidence cuts across the entire debate over intelligent design. Can there even be such a thing as evidence for an unevolved intelligence that designs biological complexity? Many naturalistic scientists and philosophers deny that there can be. But to deny it coherently, one needs an evidential framework for denying it. The preeminent framework in that regard is Bayesian. I want therefore next to examine that framework and, specifically, to show why it is inadequate both for drawing design inferences as well as for precluding them.

When the Bayesian approach tries to adjudicate between chance and design hypotheses, it treats both chance and design hypotheses as having prior probabilities and as conferring probabilities on outcomes and events. Thus, given the chance hypothesis H, the design hypothesis D and the outcome E, the Bayesian theorist attempts to compare the posterior probabilities of H and D on E (i.e., $P(H \mid E)$ vs. $P(D \mid E)$). If the posterior probability of D on E is greater than that of H on E, then E counts as evidence in favor of D, and the strength of that evidence is proportional to how much greater $P(D \mid E)$ is than $P(H \mid E)$. Unfortunately, calculating posterior probabilities requires knowing prior probabilities (i.e., $P(H)$ and $P(D)$), and often these are not available. In that case, one may merely calculate the likelihoods of E on both H and D (i.e., $P(E \mid H)$ vs. $P(E \mid D)$).

There's a stripped down version of the Bayesian approach known as the *likelihood approach* that essentially ignores prior probabilities and simply looks at the likelihood ratio (i.e., $P(E \mid H)/P(E \mid D)$) to determine the strength of evidence in favor of a hypothesis. This, however, makes for an idiosyncratic understanding of evidence. Evidence, as usually understood, refers to what causes us to revise our beliefs. But likelihood ratios are in no position to do that without help from prior probabilities. For instance, if I hear from my attic the pitter-patter of little feet and the sound of bowling pins colliding, the likelihood of the design hypothesis that gremlins are bowling in my attic may be greater than the likelihood of any chance hypothesis that purports to explain those sounds. And yet my disbelief in the gremlin hypothesis would remain as utter and complete as before because of my prior belief that gremlins don't exist. (In Bayesian terms, the prior probability $P(D)$, where D is the gremlin hypothesis, is for me effectively zero.)

I've just described the Bayesian approach to assessing the evidence for design hypotheses in comparison with chance hypotheses. Accordingly, to draw a design inference is to determine that the evidence, construed in Bayesian or likelihood terms, favors design over chance. What's wrong with this approach to inferring design? Lots. I'll briefly summarize what's wrong, bullet-point fashion. (For more details, refer to chapter two of *No Free Lunch*.)

1. Need for prior probabilities. As we've already seen, for the Bayesian approach to work requires prior probabilities. Yet prior probabilities are often impossible to justify. Unlike the example of the urn and two coins discussed earlier (in which drawing a ball from an urn neatly determines the prior probabilities regarding which coin will be tossed), for most design inferences, especially the interesting ones like whether there is design in biological systems, either we have no handle on the prior probability of a design hypothesis or that prior probability is fiercely disputed. (Theists, for instance, might regard the prior probability as high whereas atheists would regard it as low.)

2. Design hypotheses conferring probabilities. The Bayesian approach requires that design hypotheses, like chance hypotheses, confer probabilities on events. In the notation above, for the Bayesian approach to work, the likelihoods P(E | D) and P(E | H) both need to be well-defined. Suppose E denotes the event responsible for a certain gene, where this gene in turn codes for a certain enzyme. Given the various natural processes to which genes are subject (mutation, deletion, duplication, crossover, etc.), P(E | H) is well-defined. But what about P(E | D)? Assuming the enzyme in question constitutes an unprecedented biological innovation, how do we assign a probability to a designer designing it?

The difficulty here is not confined to biological design hypotheses. Indeed, it applies to all cases of innovative design. To be sure, there are design hypotheses that confer reliable probabilities. For instance, my typing this book confers a probability of about 13 percent on the letter *e*. (That's how often on average writers in English employ the letter *e*.) But what's the probability of me writing this book? What's the probability of Rachmaninoff composing his variations on a theme by Paganini? What's the probability of Shakespeare writing his sonnets? When the issue is creative innovation, the very act of expressing the likelihood P(E | D) becomes highly problematic and prejudicial. It puts creative innovation by a designer in the same boat as natural laws, requiring of design a predictability that's circumscribable in terms of probabilities. But designers are inventors of unprecedented novelty, and such creative innovation transcends all probabilities.

3. The illusion of mathematical rigor. As I noted in the previous point, if E denotes the occurrence of a certain gene coding for a certain novel enzyme, then P(E | H) can reasonably be regarded as having a well-defined probability. Provided that the problem of assessing this probability is not too technically difficult, we may be able to evaluate it precisely, or at least estimate an upper bound for it. But what about P(E | D)? What about probabilities like this more generally, where a design hypothesis confers a probability on a creative innovation? Not only is there no reason to think that such probabilities make sense (see the previous point), but when Bayesians reason with such probabilities, they do so without attaching any precise numbers to them. The probability P(E | D) functions as a placeholder for ignorance, lending an air or mathematical rigor to what really is just a subjective assessment of how plausible a design hypothesis seems to the person offering a Bayesian analysis.

4. Eliminating chance without comparison. Within the Bayesian approach, statistical evidence is inherently comparative: there's no evidence for or against a hypothesis as such but only better or worse evidence for one hypothesis in relation to another. But that all statistical reasoning should be comparative in this way cannot be right. There exist cases where one and only one statistical hypothesis is relevant and needs to be assessed. Consider, for instance a fair coin (i.e., a perfectly symmetrical rigid disk with distinguishable sides) that you yourself are tossing. If you witness a thousand heads in a row (an overwhelmingly improbable event), you'll be inclined to reject the only relevant chance hypothesis, namely, that the coin tosses are independent and identically distributed with uniform probability.

Does it matter to your rejection of this chance hypothesis whether you've formulated an alternative hypothesis? I submit it does not. To see this, ask yourself, *When do I start looking for alternative hypotheses in such scenarios?* The answer is, precisely when a wildly improbable event like a thousand heads in a row occurs. So, it's not that you started out comparing two hypotheses but rather that you started out with a single hypothesis, which, when it became problematic on account of a wild improbability (itself suggesting that Fisherian significance testing lurks here in the background), you then tacitly rejected it by inventing an alternative hypothesis. The alternative hypothesis in such scenarios is entirely ex post facto. It is invented merely to keep alive the Bayesian fiction that all statistical reasoning must be comparative.

5. Backpedaling priors. As a variant of the last point, return to the earlier example of an urn with a million balls, one black and the rest white. As before, imagine that a fair coin is to be tossed if a white ball is randomly sampled from the urn but that a biased coin with probability .9 of landing heads is to be tossed otherwise. This time, however, imagine that the coin is tossed not ten times but ten thousand times and that each time it lands heads. The probability of getting ten thousand heads in a row with the fair coin is approximately 1 in 10^{3010}; with the biased coin it is approximately 1 in 10^{458}. (With ten thousand tosses, tails are overwhelmingly likely to turn up for either coin.) A Bayesian analysis then shows that the probability that a white ball was selected is approximately 1 in 10^{2546}, and the probability that the lone black ball was selected is 1 minus that minuscule probability.

Should we therefore, as good Bayesians, conclude that the black ball was indeed selected and that the biased coin was indeed flipped? (The selection of the black ball is vastly more probable, given ten thousand heads in a row, than the selection of a white ball.) Clearly this is absurd. The probability of getting ten thousand heads in a row with either coin is vastly improbable, and it doesn't matter which urn was selected. The only sensible conclusion is that *neither* coin was randomly tossed ten thousand times. A

Bayesian may therefore want to change the prior probability to introduce some doubt about whether the urn and subsequently one of the two coins were random sampled. But as in the previous point, we need to ask what induces us to change or reevaluate our prior probabilities. The answer is not strictly Bayesian considerations but rather considerations of small probability based on chance hypotheses that, as first posed, admit no alternatives. The alternatives need, then, to be introduced subsequently because Fisherian, not Bayesian, considerations prompt them.

6. Independent empirical evidence for design. Bayesian theorists are often wedded to a Humean inductive framework in which design hypotheses require independent empirical evidence of a designer actually at work (i.e., the camera is running and the designer is—or at least in principle could be—caught on video tape) before design may be legitimately attributed. We saw in the last chapter that this restriction is not just artificial but in fact incoherent because induction cannot be the basis for identifying design, there being no way to get that induction up and running. Nevertheless, for Bayesians wedded to Hume, it is convenient to block a Bayesian analysis that might implicate design from even getting started by denying that certain design hypotheses—like a design hypothesis that appeals to an unevolved intelligence to explain biological complexity—could even in principle admit independent empirical evidence.

Thus, rather than face the problem of assessing prior probabilities in such cases, Bayesians wedded to Hume merely impose an additional restriction on the Bayesian framework by stipulating, in effect, that the Bayesian framework may not be used for design hypotheses without independent empirical evidence of a designer. Strictly speaking, this restriction has no place within the Bayesian probabilistic apparatus. (Bayes's theorem works regardless of where the probabilities associated with a design hypothesis come from. Just plug in the numbers.) But such a restriction is now increasingly being invoked against intelligent design. For instance, whereas Sober allowed considerable freedom for Bayesian design inferences in biology in his 1993 edition of *Philosophy of Biology* (and thus before intelligent design had intellectual currency), he closed off any design inference to a designer lacking independent empirical evidence in the 2000 edition of that book (after intelligent design had created considerable waves). Thus, whereas the 1993 edition gave intelligent design a lease on life, the 2000 edition took it away.

The independent empirical evidence requirement raises a curious dilemma for Darwinism. Imagine that space travelers show up loaded with unbelievably advanced technology. They tell us (in English) that they've had this technology for hundreds of millions of years and give us solid evidence of it (perhaps by pointing to some star cluster hundreds of millions of light years away whose arrangement signifies a message that confirms the aliens' claim). Moreover, they demonstrate to us that with this technology they can assemble atom by atom and molecule by molecule the most complex organisms. Suppose we have good reason to think that these aliens were here on earth at key moments in life's history (e.g., at the origin of life, the origin of eukaryotes, the origin of metazoans and the origin of the animal phyla in the Cambrian). Suppose further that in forming life from scratch the aliens would not leave any trace. (Their technology is so advanced that they clean up after themselves perfectly—no garbage or any other signs of activity would be left behind.) Suppose, finally, that none of the facts of biology are different from what they are now. Should we think that life at key moments in its history was designed?

We now have all the independent empirical evidence we could want for the existence of physically embodied designers capable of bringing about the complexity of life on earth. If, in addition, our best probabilistic analysis of the biological systems in question tells us that unguided natural processes could not have produced them with anything like a reasonable probability, is a Bayesian design inference now warranted? Could the design of life in that case become more probable than a Darwinian explanation (probabilities here being interpreted in a Bayesian or likelihood sense) simply because independent empirical evidence attests to designers with the capacity to produce biological systems?

This prospect, however, should worry Darwinists. The facts of biology, after all, have not changed. Yet design would be a better explanation if designers capable of, say, producing the animal phyla of the Cambrian could be attested through independent empirical evidence. Note that there's no smoking gun here. (We've no direct evidence of alien involvement in the fossil record, for instance.) All we know by observation is that beings with the power to generate life exist and could have acted. Would it help to know that the aliens really like building carbon-based life? But how could we know that? Do we simply take their word for it? The data of biology and natural history, we assume, stay as they are now.

But if design is a better explanation simply because of independent empirical evidence of technologically advanced space aliens, why should it not be a better explanation absent such evidence? If Darwinism is so poor an explanation that it would cave the instant space aliens capable of generating living forms in all their complexity could be independently attested, then why should it cease to be a poor explanation in the absence of those space aliens? Again, the facts of biology themselves have not changed.

Is there a way to salvage the independent empirical evidence requirement? Clearly it would be illegitimate to modify this requirement by ruling out circumstantial evidence entirely and permitting only direct "eyewitness" evidence of a designer actually manipulating the designed object in question. Even Elliott Sober would not go along with this proposal. (See his *Reconstructing the Past*. To reconstruct the past we need circumstantial evidence.) For Sober, circumstantial evidence could in principle support a biological design hypothesis. The important thing for Sober is that there be independent empirical evidence for the existence of a designer. But no smoking gun is required. In fact, to require a smoking gun in the sense of direct "eyewitness" evidence would be just as bad for Darwinism as for intelligent design. The evidence is just as circumstantial for one as for the other.

But once the independent empirical evidence for design can be circumstantial, establishing merely the existence of a designer with the causal power and opportunity to produce the effect in question (as in the alien thought experiment), we have exactly the same set of biological data to explain that we had before we acquired that evidence. The requirement for independent empirical evidence is therefore either vacuous (if it can be circumstantial) or prejudicial (if required to be direct). And in either case it obstructs inquiry into any actual design that might be present. If we require independent empirical evidence of design but don't have it, we won't see design even if it is there.

7. Implicit use of specifications. And finally we come to the most damning problem facing the Bayesian approach, namely, that it presupposes the very account of specification and rejection regions that it was meant to preclude. Bayesian theorists see specification as an incongruous and dispensable feature of design inferences. For instance, Timothy and Lydia McGrew regard specification as having no "epistemic relevance" (Symposium on Design Reasoning, Calvin College, May 2001). At that same symposium Robin Collins, also a Bayesian, remarked, "We could roughly define a specification as any type of pattern for which we have some reasons to expect an intelligent agent to produce it." Thus a Bayesian use of specification might look as follows: given some event E and a design hypothesis D, a specification would assist in inferring design for E if the probability of E conditional on D is increased by noting that E conforms to the specification (which, á la Collins, is a "pattern for which we have some reasons to expect an intelligent agent to produce it").

But there's a crucial difficulty here that Bayesians invariably sidestep. Consider the case of New Jersey election commissioner Nicholas Caputo, who was accused of rigging ballot lines. (This example appears in a number of my writings and has been widely discussed on the Internet. A ballot line is the order of candidates listed on a ballot. It is to the advantage of a candidate to be listed first on a ballot line because voters tend to vote more readily for such candidates.) Call Caputo's ballot line selections the event E. E consists of 41 selections of Democrats and Republicans in sequence with Democrats outnumbering Republicans 40 to 1. For definiteness, let's assume that Caputo's ballot line selections looked as follows. (Newspapers covering the story to my knowledge never reported the actual sequence.)

DDDDDDDDDDDDDDDDDDDDDDRDDDDDDDDDDDDDDDDDDD

Thus we suppose that for the initial 22 times, Caputo chose the Democrats to head the ballot line; then at the 23rd time, he chose the Republicans; after which, for the remaining times, he chose the Democrats.

If Democrats and Republicans were equally likely to have come up (as Caputo claimed), this event has a probability of approximately 1 in 2 trillion. This is improbable, yes, but by itself it's not enough to implicate Caputo in cheating. Highly improbable events after all happen by chance all the time. Indeed, any sequence of forty-one Democrats and Republicans whatsoever would be just as unlikely. What additionally, then, do we need to confirm cheating (and thereby design)? To implicate Caputo in cheating, it's not enough merely to note a preponderance of Democrats over Republicans in some sequence of ballot line selections. Rather, one must also note that a preponderance as extreme as this is highly unlikely. In other words, it wasn't the event E (Caputo's actual ballot line selections) whose improbability the Bayesian needed to compute but the composite event E* consisting of all possible ballot line selections that exhibit at least as many Democrats as Caputo selected. This composite event, E*, consists of 42 possible ballot line selections and has improbability 1 in 50 billion. It was this event and this improbability on which the New Jersey Supreme Court rightly focused when it deliberated whether Caputo had in fact cheated. Moreover, it's this event that the Bayesian needs to identify and whose probability the Bayesian needs to compute to perform a Bayesian analysis.

But how does the Bayesian identify this event? Let's be clear that observation never hands us composite events like E* but only elementary outcomes like E (i.e., Caputo's actual ballot line selection and not the ensemble of ballot line selections as extreme as Caputo's). But from whence came this composite event? Within the Fisherian framework the answer is clear: E* is the rejection region (and therefore the specification) that counts the number of Democrats selected in 41 tries and totals at least as many Democrats as in Caputo's ballot line selections. That's what the court used, and that's what Bayesians use. Bayesians, however, offer no account of how they identify the events to which they assign probabilities. If the only events they ever considered were elementary outcomes, there would be no problem. But that's not the case. Bayesians routinely consider such composite events. In the case of Bayesian design inferences (and Bayesians definitely want to draw a design inference with regard to Caputo's ballot line selections), those composite events are given by specifications.

Let me paint the picture more starkly. Consider an elementary outcome E. Suppose initially we see no pattern that gives us reason to expect an intelligent agent produced it. But then, rummaging through our background knowledge, we suddenly see a pattern that signifies design in E. Under a Bayesian analysis, the probability of E given the design hypothesis suddenly jumps way up. That, however, isn't enough to allow us to infer design. As is usual in the Bayesian scheme, we need to compare a probability conditional on design with one conditional on chance. But for which event do we compute these probabilities? As it turns out, we do so not for the elementary outcome E but for the composite event E* consisting of all elementary outcomes that exhibit the pattern signifying design. Indeed, it does no good to argue for E being the result of design on the basis of some pattern unless the entire collection of elementary outcomes that exhibit that pattern is itself improbable on the chance hypothesis. The Bayesian therefore needs to compare the probability of E* conditional on the design hypothesis with the probability of E* conditional on the chance hypothesis.

The bottom line is this: the Bayesian approach to statistical rationality is parasitic on the Fisherian approach and can properly adjudicate only among hypotheses that the Fisherian approach has thus far failed to eliminate. In particular, the Bayesian approach offers no account of how it arrives at the events on which it performs a Bayesian analysis. The selection of those events is highly intentional, and in the case of Bayesian design inferences it needs to presuppose an account of specification. Specified complexity, far from being refuted by the Bayesian approach, is therefore implicit throughout Bayesian design inferences.

To sum up, there is no merit whatsoever to the charge that by looking to specified complexity to infer design, intelligent design violates statistical rationality. Quite the contrary. By developing specified complexity as an analytic tool for inferring design, intelligent design advances the study of scientific reasoning and vindicates the Fisherian approach to statistical rationality.

34

THE DEMAND FOR DETAILS: DARWINISM'S *TU QUOQUE*

Isn't it the height of hypocrisy for design theorists to complain that Darwinism provides no details about the emergence of biological complexity when their own theory, intelligent design, likewise provides no such details?

ACCORDING TO DARWINISTS, neither specified complexity nor irreducible complexity is beyond the reach of the Darwinian mechanism. Yet to justify this claim, all that Darwinists do is describe, in highly abstract and schematic terms, supposedly possible Darwinian pathways that might bring about those features of living systems. The proposed evolutionary pathways are all highly imaginative and speculative. But where are the details? No Darwinist, for instance, has offered a hypothetical Darwinian production of any tightly integrated multipart system with enough detail to make the hypothesis testable even in principle.

Intelligent design challenges Darwinism, therefore, by demanding details. Design proponents want to see the details by which the Darwinian mechanism is supposed to accomplish the magnificent feats attributed to it. But since those details are neither available nor forthcoming, design proponents suggest that it is time for Darwinism's exclusion of design from biology to end. Darwinists, to be sure, will have none of this. In particular, they don't regard the absence of such details as reason to question Darwinism. Yet by neither filling in those details themselves nor pointing us to where they may be found, they do not help Darwinism's credibility. To be sure, convinced Darwinists need no such details, but skeptical outsiders do, and those outsiders constitute the bulk of the American population (hence the ongoing controversy over the teaching of biological evolution in Kansas, Ohio, Cobb County, etc.).

Darwinists therefore turns the tables with a *tu quoque*. Consider, for instance, the following remark by H. Allen Orr in his review of my book *No Free Lunch* for the *Boston Review* (<bostonreview.net/BR27.3/orr.html >):

> The causal specificity argument [i.e., demand-for-details argument] is also an exercise in nerve. We are, recall, trying to choose between two theories. One says bacterial flagella were built by mutation and selection and the other says they were built by an intelligent designer. And Dembski concludes the *first* theory lacks historical concreteness? *Darwinism* suffers a shortage of specificity? When, after all, did Dembski's designer come up with plans for flagella? Just how did he reach out and shape that flagellum? Which protein did he move first or did he touch them all at once? It is the height of hypocrisy for Dembski to complain that Darwinism lacks causal specificity when his own theory lacks *any* specificity, including one atom of historical concreteness. Dembski may not have much of an argument, but you've got to admit he's got chutzpah.

Orr's *tu quoque* is not nearly as potent as he thinks. That's because the demand for details is a burden for Darwinism in a way that it is not for intelligent design. Darwinism is a theory about *process*. Darwinism says that a certain type of process took organisms of type A and transformed them into organisms of type B. The Darwinian process occurs in discrete steps (the finest level of resolution of those steps being the generation of one organism from another in reproduction). Darwinism is committed to a sequence of manageable steps that gradually transforms A into B. In consequence, there has to be some sequence such that $A = A_1$ transforms into A_2, which in turn transforms into A_3, . . . which then transforms into $A_n = B$, where each transition from one step to the next can readily be accounted for in terms of natural selection and random variation. Thus, for instance, in a Darwinian explanation of the bacterial flagellum, we know that bacteria lacking a flagellum (and also lacking any genes coding for a flagellum) had to evolve into bacteria with a flagellum (and thus possessing a novel genetic complement for the flagellum). If Darwinism is correct, some step-by-step Darwinian process had to take us from the former type of bacteria to the latter. So how did it happen? How could it have happened? Nature somehow filled in the details, but Darwinists never do. This is a fault in Darwin's theory, and intelligent design is rightly drawing attention to it.

But what about intelligent design? Darwinists suggest that the same fault applies to it, but it does not. Intelligent design, in contrast to Darwinism, is not a theory about process but about *creative innovation*. Now creative innovation is not a process. Creative innovation can occur in a process, but even then it is a process where each step constitutes an individual creative act (a micro-innovation, as it were). In our experience with intelligences, creative innovation is a unifying conceptual act that ties together disparate elements into a purposeful whole. The act can occur over time in a process, or it can occur in one fell swoop. But in either case, creative innovation is not reducible to a causal chain where one step "causes" the next. The demand for details is a demand for causal specificity: that is, it is about finding the precise causal antecedents that account for and thus predict (whether deterministically or probabilistically) an event, object or structure. But intelligences are free. In the act of creation they violate expectations. They create as they choose to create. There's nothing that required Mozart to compose his Jupiter Symphony or Bell to invent the telephone or Shakespeare to write *King Lear*. And there's no way to have predicted these creative innovations. Consequently, the demand for causal details applies secondarily, not primarily, to creative innovation and therefore to intelligent design.

In the theory of intelligent design, the demand for causal details comes up in the antecedent circumstances that condition (but do not determine, explain or account for) creative innovation. Antecedent circumstances set the stage for creative innovation. Technologies, for instance, evolve by building on previous technologies. But they evolve in the first instance by inventors having ideas. Where do those ideas come from? Antecedent circumstances are not much help here. No set of antecedent circumstances can account for a creative innovation. Antecedent circumstances, however, need definitely to be considered for their effect on constraining the innovations that are produced. Beethoven, for instance, could not have written music for the piano until after the piano was invented. The philosopher Georg Hegel generalizes this point beautifully with regard to the unfolding of culture, showing how one cultural advance builds on the next. But note that for Hegel it is intelligence *(Geist)* that does the building and not brute material processes.

In the case of intelligent design, the demand for details certainly applies to the antecedent circumstances that lead up to a creative innovation. It also applies to the aftermath of a creative innovation. Creative innovations, after all, have consequences. Causal chains flow up to and out from them. Indeed, that is where much of the intellectual labor on intelligent design will focus in coming years, namely, in tracing the antecedent circumstances that lead up to and thereby condition the design of biological systems and then in tracing the impact of those systems throughout the biological world. The demand for details therefore remains a live issue for intelligent design. But it is not the primary issue. The primary issue is to determine whether there is design (i.e., creative innovation by an intelligence) in the first place. And for that you need specified complexity.

Bottom line: Darwinism has a burden of proof that intelligent design does not have. Darwinism is a theory of process and therefore needs to provide convincing evidence that the processes it describes are able to bear the weight placed on them. That weight is considerable—indeed, no less than the whole of biological complexity and diversity. Intelligent design by contrast has a different burden. As a theory of creative innovation, its burden is to show where creative innovations first emerge and then to trace their causal antecedents and consequents. (Note that work in this area is ongoing. See chapter forty-three.) Darwinism and intelligent design therefore face fundamentally different tasks.

35

DISPLACEMENT AND THE NO FREE LUNCH PRINCIPLE

How do the No Free Lunch theorems undercut Darwinian theory and support intelligent design?

GIVEN THE TITLE OF MY BOOK *No Free Lunch*, it's not surprising that critics see it as depending crucially on the No Free Lunch (NFL) theorems of David Wolpert and William Macready. And indeed, the title was meant to allude to those theorems. But my actual use of the NFL theorems has always been quite limited, and I certainly never argued that the NFL theorems provide a direct refutation of Darwinism or a direct confirmation of intelligent design. In fact, my key point in appealing to the NFL theorems concerns not those theorems themselves but a more general result that I call *displacement*. The NFL theorems constitute a particular instance of displacement.

The basic idea behind displacement is this: Suppose you need to search a space of possibilities. The space is so large and the possibilities individually so improbable that an exhaustive search is not feasible and a random search is highly unlikely to conclude the search successfully. As a consequence, you need some constraints on the search, some information to help guide the search to a solution. (Think of an Easter egg hunt where no one provides hints or guidance versus one where someone directs you by saying "warm," "warmer" and "hot.") All such information that assists your search, however, resides in a search space of its own—an informational space. So the search of the original space gets displaced to a search of an informational space in which the crucial information that constrains the search of the original space resides. I then argue that this higher-order informational space ("higher" with respect to the original search space) is always at least as big and at least as hard to search as the original space. I call this the *displacement problem*.

Think of it this way. Imagine an island with buried treasure. You can scour the island trying to find the buried treasure. Alternatively, you can try to find a map that tells you where the treasure is buried. Once such a map is in hand, finding the treasure is no problem. But how to find such a map? Suppose such a map exists but is mixed among a huge assortment of other maps. Finding the right map within that huge assortment will then be no easier than simply searching the island directly. The huge assortment of maps is the informational space associated with the original search space. In general, an informational space is no easier to search than the original search space.

It follows that constraining the search of an original space by employing information does not provide a nonteleological, design-free explanation for the success of that search (if indeed it turns out to be successful). Instead, the solution found in the original space merely reflects the solution already in hand in a higher-order informational space. And if the one solution exhibits specified complexity, then so does the other. In particular, when nontelic processes output specified complexity, it is because they take preexisting specified complexity and merely re-express it. They are not generating it for free or from scratch. To claim otherwise is like filling one hole by digging another. If the problem was to be rid of holes (i.e., design) completely, then the problem hasn't been resolved but merely relocated.

It's against this backdrop of displacement that I treat the NFL theorems. These theorems say that when averaged across all fitness measures from a given class, no evolutionary algorithm is superior to blind or random search. Think of a fitness measure as assigning to each item in the original search space a numerical degree of fitness. (Zero fitness here signifies minimal fitness; there's no bound on how large fitness can become.) Each fitness measure therefore constitutes an item of information that constrains an otherwise unconstrained search in the original search space.

The NFL theorems come in a variety of forms depending on the class of fitness measures being averaged over. The two theorems originally published by Wolpert and Macready ("No Free Lunch Theorems for Optimization," 1996, available at <citeseer.nj.nec.com/wolpert96no.html>) considered all fitness measures on a given space (theorem 1) as well as all fitness measures indexed by time on a given space (theorem 2). Note that with both of these theorems, fitness measures of a given type were left wholly unconstrained: all fitness measures of a given type (nonindexed in theorem 1, time-indexed in theorem 2) were there. What's more, the evolutionary algorithms employing these fitness measures were "no prior knowledge" algorithms. "No prior knowledge" simply means that the algorithm has no additional information for finding a solution other than what it gets from the fitness measures. In general, arbitrary, unconstrained, maximal classes of fitness measures each seem to have a No Free Lunch theorem for which evolutionary algorithms cannot, on average, outperform blind search.

The obvious way to try to get around NFL is to start constraining fitness measures. Say you don't like time-dependent fitness measures that vary independently of the progress of the evolutionary algorithm in reaching a solution; then constrain this class of fitness measures so that they depend on progress to a solution. Say you don't like fitness measures closed under permutation (see Christian Igel and Marc Toussaint, "On Classes of Functions for Which No Free Lunch Results Hold," at <citeseer.nj.nec.com/528857.html>); then focus on classes that are not closed under permutation. All such focusing and constraining, however, imparts information. Provided that information is both complex and specified, I show in *No Free Lunch* that such evolutionary processes never output more specified complexity than was programmed into them through such constraining of the underlying information space.

To try to get around NFL and the displacement problem more generally, Darwinists start by noting that fitness in biology varies with time. As organisms evolve and the environment changes, what the environment deems fit changes as well. But what exactly constrains the transition from one fitness measure to the next? If there is no constraint, then we are in the position of Wolpert and Macready's theorem 2, with evolutionary algorithms proceeding independently of their progress to solution and thus unable to outperform blind search. Conveniently, Darwinists never tell us what constrains the transitions. Presumably nature, unprogrammed and unguided, spontaneously gives rise to the right and needed transitions between successive fitness measures, thereby ensuring a form of complexity-increasing evolution. But that is precisely what needs to be explained.

Just try to program time-varying/coevolving fitness measures and see whether they produce solutions to interesting problems (i.e., produce specified complexity). You'll find one of two things. Either you'll get sludge because you didn't adequately constrain how fitness landscapes vary with time in response to a changing environment, or you'll get something interesting (specified complexity) because you carefully introduced constraints and thereby did design work that cannot be reduced to material mechanisms. Darwinists often give the impression that it's enough simply to dig up some replicators, place them in an environment and let the Darwinian mechanism run its course. *Presto!* Nifty things will automatically happen. No need to ask for details about how these nifty things happen because the alternative (intelligent design) is unthinkable. But in fact, nifty things do not automatically happen unless the Darwinian mechanism is itself suitably programmed, as the literature on evolutionary computation is now making clear. (See, for instance, John Bracht's "Inventions, Algorithms and Biological Design" at <www.iscid.org/papers/Bracht_InventionsAlgorithms_112601.pdf>.)

If Darwinism is going to be a theory that accounts for how novel biological information is generated, it cannot simply be a theory about who gets to survive and reproduce and who doesn't. In other words, it cannot simply be a theory about "sheer cold demographics," as Rochester University biologist Allen Orr claims. There are environments with replicators that operate strictly under the control of a Darwinian mechanism and that never do anything interesting; that is, they never exhibit a net increase in specified complexity. I detail several such scenarios in *No Free Lunch,* notably Sol Spiegelman's experiment on the evolution of polynucleotides in a replicase environment. Here a purely Darwinian form of evolution ran along and actually simplified the replicators to make them as efficient as possible in their replication. To be sure, none of this contradicts Darwinism. But that's just the point: because "sheer cold demographics" is compatible with evolution leading to no net increase in specified complexity and quite possibly to a net decrease, something else besides Darwinism must account for why we see a net increase of specified complexity over the course of biological evolution in the real world.

Simplicity by definition always entails a lower cost in raw materials than increases in complexity, and so there is an inherent tendency in evolving systems for selection pressures to force such systems toward simplicity. This is not to say that Darwinism requires or entails that evolution proceed toward simplicity. The point is simply that Darwinism, in itself, does not mandate increasing complexity and inherently favors simplicity. Thus, if we see increasing complexity, something besides Darwinism must be at work. Now Darwinists have offered several rationales for why we should expect increasing complexity strictly on Darwinian grounds (e.g., the irreversibility of certain changes and Stephen Jay Gould's lower wall of complexity below which things are dead). But all of these rationales are post hoc; in each case the opposite might well have happened and Darwinism would still be true. Thus we can imagine (and even program on a computer) Darwinian evolutionary scenarios in which reversibility has a selective advantage, in which arms races are won by simplifying and in which the lower wall of complexity is an absorbing barrier where maximal fitness is conferred by being maximally simple.

According to Stuart Kauffman, it is a deep problem why the universe is complexity-increasing (ISCID online chat November 15, 2002, available at <www.iscid.org/stuartkauffman-chat.php>):

> One of the deep puzzles is why the universe has become complex. Why has the biosphere become complex? Why has the number of ways of earning a living increased so dramatically? We have no theory about this overwhelming feature of our universe.

What for Darwinists is a foregone conclusion is for Kauffman, who is not a Darwinist, a mystery. Who's right? Is the fact that the biophysical universe is complexity-increasing a deep problem awaiting resolution, or is it one that, at least in biology, has received a decisive answer from Darwinism? Clearly, if Darwinism provided a decisive answer, someone of Kauffman's caliber would not be proclaiming the problem a mystery. And indeed, the Darwinian literature shows a complete absence of non-question-begging solutions to this problem.

In concluding this chapter, I want to draw an analogy between displacement and Church's thesis. Church's thesis is a deep claim from mathematical logic about the nature of computation. Displacement, or what I also call the *No Free Lunch principle* (not to be confused with the No Free Lunch theorems), functions within the theory of intelligent design much as Church's thesis functions within the theory of computation. Church's thesis states that if you have some procedure that is intuitively computable (i.e., that can be characterized with well-defined rules), then it can be coded as an algorithm running on a Turing machine. The No Free Lunch principle states that if you have some naturalistic process whose output exhibits specified complexity, then that process was front-loaded with specified complexity. The task of the design theorist in that case is to "follow the information trail" and show where the outputted specified complexity was first inputted (much as the task of the computer scientist is to show how some procedure that is intuitively computable can be explicitly formulated as an algorithm capable of being run on a Turing machine).

Note that there is no—and indeed can be no—mathematical proof in the strict sense of either Church's thesis or the No Free Lunch principle. (For the latter, there's always the outside possibility that specified complexity could happen as a wildly improbable event.) Even so, the two are subject to empirical verification. With Church's thesis, the challenge is to show that the intuitively computable invariably submits to the formally computable (which it has). With the No Free Lunch principle, the challenge is to show where the specified complexity outputted by a naturalistic process was in fact front-loaded.

Evolutionary biologists regularly claim to obtain specified complexity for free or from scratch. (Richard Dawkins and Thomas Schneider are some of the worst offenders in this regard.) The No Free Lunch principle counsels us to find where specified complexity supposedly gotten for free has in fact been front-loaded, smuggled in or hidden from view. Darwinism is an exercise in creative bookkeeping for hiding that its explanatory debts far exceed its explanatory resources. Think of the No Free Lunch principle, therefore, as an auditor's tool for scrutinizing Darwinism's inflated claims and showing its debts to be in default. Fortunately, as recent corporate debacles have taught us, creative bookkeeping can at best postpone but not avoid an official declaration of bankruptcy.

36

THE ONLY GAMES IN TOWN

*Isn't it crude and simplistic to cast the
debate over biological evolution as merely
between Darwinism and intelligent
design? Surely evolutionary biology
encourages many more options.*

IN THE MARCH 2003 ISSUE OF *Commentary*, design critic Paul Gross responded to design sympathizer David Berlinski for an article Berlinski had written in the December 2002 issue of *Commentary*. In that article, titled "Has Darwin Met His Match?" Berlinski characterized the controversy over biological evolution as one between Darwinism and intelligent design. In his response, Gross took Berlinski to task not just for characterizing the controversy this way (Darwinism versus intelligent design), but for even mentioning the term *Darwinism* at all. According to Gross, only "those who do not know much evolutionary biology" refer to something called "Darwinism."

Evolutionary biology, we are assured, is far richer than the caricature of it called Darwinism. I have faced this charge as well in giving public lectures critical of Darwinism. I want therefore to clarify why design theorists are not merely attacking a strawman when they criticize Darwinism and why they are not merely posing a false dilemma when they set intelligent design in opposition to Darwinism. To see this, let's begin with Berlinski's reply to Gross in the March 2003 issue of *Commentary*:

> The professionals know better. I quite understand Mr. Gross's concern. The term "Darwinism" conveys the suggestion of a secular ideology, a global system of belief. So it does and so it surely is. Darwin's theory has been variously used—by Darwinian biologists—to explain the development of a bipedal gait, the tendency to laugh when amused, obesity, *anorexia nervosa*, business negotiations, a preference

for tropical landscapes, the evolutionary roots of political rhetoric, maternal love, infanticide, clan formation, marriage, divorce, certain comical sounds, funeral rites, the formation of regular verb forms, altruism, homosexuality, feminism, greed, romantic love, jealousy, warfare, monogamy, polygamy, adultery, the fact that men are pigs, recursion, sexual display, abstract art, and religious beliefs of every description. . . .

I am also hardly the only one to use the term "Darwinism" and so convey the suggestion of an ideological agenda. Adding his mite to D. S. Bendall's collection, *Evolution from Molecules to Men* (1983), Richard Dawkins entitled his essay "Universal Darwinism." Dawkins liked the word well enough to use it again in "Darwin and Darwinism," the title of his contribution to Microsoft's *Encarta Encyclopedia*. Then there is the series of short books appearing under the title *Darwinism Today* and published by Yale University Press. The first book in the series is by the eminent Darwinian biologist John Maynard Smith.

It is no accident that in debates over biological evolution Charles Darwin's name and theory keep coming up. Nor are repeated references to Darwin and Darwinism simply out of respect for the history of the subject, as though evolutionary biology needed constantly to be reminded of its founding father. Darwin looms larger than life in the study of biological origins because his theory constitutes the very core of evolutionary biology. Indeed, nothing in evolutionary biology makes sense apart from Darwinism.

To see this, we need to understand what makes evolutionary biology tick. Although evolutionary biology is committed to common descent (i.e., that all organisms trace their lineage back to some last universal common ancestor), that is not its central claim. Indeed, there are design theorists who cheerfully hold to common descent (e.g., Michael Behe). The central claim of evolutionary biology, rather, is that an unguided physical process is sufficient to account for the emergence of biological complexity and diversity. Filling in the details of that process remains a subject for debate among evolutionary biologists. Yet it is an in-house debate and one essentially about details. In broad strokes, however, any unguided physical process capable of producing biological complexity will have to consist of three components: hereditary transmission, incidental change and natural selection.

The picture is this: We have some organism. It incurs some change. The change is incidental in the sense that it doesn't anticipate future changes that subsequent generations of organisms may experience. What's more, the change is heritable and therefore can be transmitted to the next generation. Whether it actually is transmitted to the next generation and then preferentially preserved in subsequent generations, however, depends on whether the change is in some sense beneficial to the organism (or, more obliquely, whether it is associated with other changes that are beneficial to the organism). If so, then natural selection will be likely to preserve organisms exhibiting that change.

This picture is perfectly general. It can accommodate hereditary transmissions that are genetic as well as those that are epigenetic. (Epigenetic hereditary transmissions leave genetic information intact. Much of biology these days is committed to genetic heritability, though there is now increasing evidence that epigenetic factors play a role in hereditary transmission.) It can accommodate Lamarckian evolution, whose incidental changes occur as organisms, simply by putting to use existing structures, or enhance or modify the functionalities of those structures. It can also accommodate all forms of what typically is meant by Darwinian evolution, including neo-Darwinism, whose incidental changes are accidental errors in the genome. It can as well accommodate Lynn Margulis's idea of symbiogenetic evolution, whose incidental changes occur as different types of organisms come together to form a new, hybrid organism. It can even accommodate self-organizational forms of evolution, whose incidental changes result from self-organizational processes. (Note that precisely because what's doing the organizing is the self—the individual organism—self-organizational processes must operate within a given generation and cannot operate cross-generationally.) Other forms of incidental change include genetic drift, lateral gene transfer and the action of regulatory genes in development.

Evolutionary biologists debate the precise role and extent of hereditary transmission and incidental change. The debate can even be quite sharp at times. See, for instance, the introduction to Margulis's book *Acquiring Genomes*, in which she slams neo-Darwinism for having the wrong picture of incidental change. (According to her, symbiogenetic change, and not random genetic mutation, is the primary form of incidental change that drives evolution.) Yet Ernst Mayr, one of the architects of the neo-Darwinian synthesis of the 1930s and 1940s, writes a glowing foreword to Margulis's book. Why should a proponent of neo-Darwinism write a foreword to a book that is highly critical of neo-Darwinism? The answer is that Margulis doesn't challenge Darwinism's holy of holies—natural selection. Indeed, at heart she is just as much a Darwinist as Mayr. Darwin himself was unclear about the mechanisms of hereditary transmission and incidental change. But whatever form they took, Darwin was convinced that natural selection was the key to harnessing them. Ditto for Mayr. Ditto for Margulis. That's why to this day we hear repeated references to Darwin's theory of natural selection but not to Darwin's theory of variation or Darwin's theory of inheritance.

Indeed, without intelligent design, what else besides natural selection can even pretend to coordinate the incidental changes that hereditary transmission passes from one generation to the next? Incidental changes by themselves cannot be responsible for the massive increases in biological complexity that have occurred in natural history. Even if one includes symbiogenesis, lateral gene transfer, the action of regulatory genes in development and other processes capable of inducing significant evolutionary change in a single generation, those changes would still need to be cumulated and coordinated over successive generations. Apart from intelligent design, there's only one natural process that could even in principle cumulate and coordinate such changes—natural selection.

That's why Stuart Kauffman will write, in *At Home in the Universe,*

> Biologists now tend to believe profoundly that natural selection is the invisible hand that crafts well-wrought forms. It may be an overstatement to claim that biologists view selection as the sole source of order in biology, but not by much. If current biology has a central canon, you have now heard it.

Kauffman, though critical of Darwinian evolutionary theory and skeptical of natural selection as the great coordinator of incidental change in biology, nonetheless has no naturalistic alternative to natural selection.

That's also why Alvin Plantinga has remarked that if materialism is true, then Darwinism is "the only game in town" (February 22, 1997, University of Texas at Austin, at a conference titled "Naturalism, Theism and the Scientific Enterprise"). Phillip Johnson elaborates:

> If materialism is true, then something like Darwinism also has to be true, regardless of the evidence. Materialism requires that chemicals must have the capacity to form living organisms, and that a primal Replicator must be able to evolve all the complex features of plants and animals without the aid of a Designer. So evolution must be a mindless process that starts with chance (mutation), and employs something capable of designing complicated structures (natural selection). That's Darwinism, and if it isn't true, then the materialist project

lacks a creation story. ("Dogmatic Materialism," *Boston Review,* February 1997; <http://bostonreview.net/br22.1/johnson.html >)

In short, evolutionary biology needs a designer substitute to coordinate the incidental changes that hereditary transmission passes from one generation to the next, and there's only one naturalistic candidate on the table, to wit, natural selection. Indeed, it's no accident that the word *selection* and the word *intelligence* are etymologically related—the *lec* in *selection* has the same root as the *lig* in *intelligence*. Both derive from the same Indo-European root meaning "to gather" and therefore "to choose." Before Darwin, the ability to choose was largely confined to designing intelligences, that is, to conscious agents that could reflect deliberatively on the possible consequences of their choices. Darwin's claim to fame was to argue that natural forces, lacking any purposiveness or prevision of future possibilities, likewise have the power to choose via natural selection.

In ascribing the power to choose to unintelligent natural forces, Darwin perpetrated the greatest intellectual swindle in the history of ideas. Nature has no power to choose. All natural selection does is narrow the variability of incidental change by weeding out the less fit. What's more, it acts on the spur of the moment, based solely on what the environment at the present time deems fit and thus without any prevision of future possibilities. And yet this blind process, when coupled with another blind process, namely, incidental change, is supposed to produce designs that exceed the capacities of any designers in our experience. No wonder Daniel Dennett, in *Darwin's Dangerous Idea*, credits Darwin with "the single best idea anyone has ever had." Getting design without a designer is a good trick indeed. Darwin was like a magician performing far enough away from his subjects that he could dazzle them—until somebody starts handing out binoculars. Darwin's idea was a good trick while it lasted. But with advances in technology as well as the information and life sciences (especially molecular biology), the Darwinian magic gig is now up. It's time to lay aside the tricks—the smokescreens and the handwaving, the just-so stories and the stonewalling, the bluster and the bluffing—and to explain scientifically what people have known all along, namely, why you can't get design without a designer. That's where intelligent design comes in.

According to Francisco Ayala, "It was Darwin's greatest accomplishment to show that the directive organization of living beings can be explained as the result of a natural process, natural selection, without any need to resort to a Creator or other external agent" (quoted from Ayala's essay "Darwin's Revolution" in *Creative Evolution?!*). In thus describing

Darwin's main claim to fame, Ayala hit the nail on the head. A designer might resort to bringing about the diversity of life by means of the Darwinian mechanism, but the Darwinian mechanism need not resort to a designer to bring about the diversity of life. It follows that the Darwinian mechanism does not, and indeed cannot, make manifest any design that a designer might have placed in the world by means of it. Granted, the design might still be there, but it would not be discernible.

Design can be laid on top of Darwin's theory, but it cannot properly be integrated with it. To see this, consider the following claim put forth by Kenneth Miller in *Finding Darwin's God:*

> The indeterminate nature of quantum events would allow a clever and subtle God to influence events in ways that are profound, but scientifically undetectable to us. Those events could include the appearance of mutations, the activation of individual neurons in the brain, and even the survival of individual cells and organisms affected by the chance processes of radioactive decay.

Miller here interprets quantum indeterminacy as an occasion for divine action (and, by implication, design). Nevertheless, he makes sure his interpretation changes no facts or theories of science. Accordingly, God's action and, more generally, the activity of a designing intelligence, must be scientifically undetectable and must remain scientifically undetectable in biology. Otherwise, Darwin's theory would become a teleological theory with random variations giving way to guided variations whose guidance is scientifically detectable. (In other words, the changes that form the raw material for evolution would no longer be incidental.)

Among the theological cognoscenti, Miller's "God of the quantum gaps" is regarded as crude and unnecessary. The preferred method for understanding divine action is to distinguish primary from secondary causation, leaving primary causation to philosophy and theology and secondary causation to science. Accordingly, God, the primary cause, employs secondary causes, such as the ordinary processes of physics and chemistry, to accomplish God's purposes. Given the distinction between primary and secondary causes, Miller's model of divine action at quantum interstices becomes an exercise in irrelevance. Science treats secondary causes on their own terms, without reference to God's purposes and thus without any need for God secretly to intervene in the causal processes of nature. Consequently, science has no purchase and need seek no purchase on divine action or any other form of design. In general, the distinction between primary and secondary causes renders divine action both invisible and irrelevant to science. This distinction is ultimately what's behind such popular strategies for making peace between science and religion such as Stephen Jay Gould's NOMA (nonoverlapping magisteria), Howard Van Till's RFEP (robust formational economy principle) and the complementarianism that pervades much of the mainstream science-theology discussion.

All such maneuvers by theistic evolutionists to bring divine action into consonance with science leave the content of science, and thus of Darwinian evolutionary theory, untouched. Thus, when these maneuvers are used to attribute design to features of the world, they do so in spite of the science and not because of it. From the vantage of Darwinian evolutionary theory, the emergence of biological complexity and diversity is as much to be expected as the emergence of twenty heads in a row among a crowd of a million coin tossers where each participant must exit the game when he or she tosses tails. (Imagine each person in the crowd tosses a coin and keeps standing so long as he or she tosses heads but must sit down otherwise, thereby exiting the game.)

The science of coin tossing (probability theory) tells us that out of a million such coin-tossers one person will on average be left standing who has tossed twenty heads in a row. So too the science of Darwinian evolution tells us that periodically nature will produce beneficial incidental changes and, rather than simply disappearing into the dust of prehistory, these changes will be passed on by the power of natural selection. Through a long chain of such events, the cumulative effect of natural selection and incidental change over several billion years is likely to produce the degree of biological complexity and diversity we observe now. Just as the science of coin tossing does not justify attributing to the person who tossed twenty heads in a row any special skill or wisdom at coin tossing, so too the science of Darwinian evolution does not justify attributing to the evolutionary process any special skill or wisdom at generating biological complexity and diversity. In each case the outcome is properly regarded as expected or predictable and not as a creative achievement that is the result of design.

Darwinism, though not excluding design on metaphysical or theological grounds, nonetheless renders design undetectable to science. What's more, so long as design is undetectable to science, Darwinian evolution—broadly construed in terms of hereditary transmission, incidental change and natural selection—is the only scientific possibility. Or, as we said earlier, if materialism is true, then Darwinism is the only game in town, scientifically speaking. It follows that the only scientific alternative to Darwinism is a theory that can detect design in nature and that succeeds in doing so for biological systems. How could this happen? In detecting design, such a theory must at minimum show that the components of Darwinian evolution—hereditary transmission, incidental change and natural selection—have no inherent capacity to produce certain features of biological systems. (Otherwise, there would be no reason to question Darwinism, the presumption being that natural selection has acquitted itself admirably as a designer substitute.)

What, then, would such features beyond the reach of Darwinian evolution look like? Certainly they would need to be highly improbable with respect to Darwinian evolution, for otherwise Darwinian evolution could readily account for them. But sheer improbability is not enough inasmuch as highly improbable things happen by chance all the time (see chapter eight). As a consequence, for such features to reside beyond the reach of Darwinian evolution, they also need to be suitably patterned, or specified (see chapter ten). But for something to be highly improbable and specified means that it exhibits specified complexity; and specified complexity is a reliable empirical marker of actual design (see chapter twelve). Thus we see that as soon as the possibility of a scientific alternative to Darwinism is raised, logic leads us inescapably to a theory of intelligent design with specified complexity at its center. Either all features of biological systems result from hereditary transmission, incidental change and natural selection, or there are some features that exhibit specified complexity and therefore also result from design.

It follows that Darwinism is not the only game in town. For Darwinism to be the only game in town would require materialism to be inescapably true. But materialism is not the conclusion of a valid scientific inference or an inescapable truth of reason. Rather, it is an ideology that increasingly suffocates scientific inquiry. The refusal of Darwinists to acknowledge intelligent design speaks no more to its truth and validity than the segregationist policies of the days before the civil rights movement spoke to the dignity and ability of minorities being ostracized. How many games are there in town? The town here is science, and the game objective is to explain the emergence of biological complexity and diversity. Despite Darwinist denials, there are in fact two games in town, two scientific theories for explaining biological origins: Darwinism and intelligent design. Remarkably, these are the only games in town.

PART 6

A NEW KIND
OF SCIENCE

37

ASPIRATIONS

*What does science stand to gain from
intelligent design, and what is intelligent
design aspiring to do for science?*

SCIENCE IS A VAST AND VARIED ENTERPRISE. Some aspects of science have
been hugely successful. Others have not. Proponents of intelligent design contend that biology as currently practiced has been hugely unsuccessful at resolving the problem of life's origin and the subsequent emergence of biological complexity. Does the success of science as a whole
warrant that biologists stick to their guns and resist the incursions of design and teleology? The challenge of intelligent design is real. This is not
like someone claiming that ancient technologies could not have built the
pyramids, so gods or goblins must have done it. We can show how, with
the technological resources at hand, the ancient Egyptians could have
produced the pyramids.

By contrast, material mechanisms known to date offer no such insight
into biological complexity. Cell biologist Franklin Harold in *The Way of the
Cell* (Oxford, 2001) remarks that in trying to account for biological complexity, biologists thus far have merely proposed "a variety of wishful
speculations." If biologists really understood the emergence of biological
complexity in purely material terms, intelligent design couldn't even get
off the ground. The fact that they don't accounts for intelligent design's
quick rise in public consciousness. Show us detailed, testable, mechanistic
models for the origin of life, the origin of the genetic code, the origin of
ubiquitous biomacromolecules and assemblages like the ribosome, and
the origin of molecular machines like the bacterial flagellum, and intelligent design will die a quick death.

But that hasn't happened. Nor does it show any sign of happening. As
my colleague Robert Koons points out, in attempting to account for the
emergence of biological complexity, all evolutionists have done is de-

scribe supposedly possible mechanisms, in highly abstract and schematic terms, to which, in the case of Darwinism, no significant details have been added since the time of Darwin (and, one can argue, none has been added even since the time of Empedocles and Epicurus two thousand years earlier) and for which other naturalistic evolutionary scenarios remain even more speculative.

Critics of evolution who say it is merely a theory don't go far enough. It doesn't even deserve to be called a theory—at least not when purporting to account for the emergence of biological complexity. No Darwinist, for instance, has offered a hypothetical Darwinian production of any tightly integrated multipart "adaptation" with enough specificity to make the hypothesis testable even in principle. In accounting for biological complexity, evolutionary biology isn't so much a theory as a pile of promissory notes for future theories, none of which has been redeemed since the publication of Darwin's *Origin of Species* almost 150 years ago. Moreover, when these promissory notes are called into question, as intelligent design is increasingly doing, Darwinism becomes an exercise in rationalizing why these promissory notes should not be regarded as in default. Remove the promissory notes and rationalizations, and Darwinism becomes a quite modest theory.

Even with this remarkably bad track record, evolutionary biologists are reluctant to give up on Darwinian and other materialistic accounts of biological complexity. For many scientists, a remarkably bad theory—even a pile of promissory notes and rationalizations—is better than nothing. Before jumping ship, scientists want a positive alternative. Intelligent design offers just that, providing a positive design-theoretic alternative to the proposed materialist accounts of biological complexity. To be fully successful at overturning its materialistic alternatives, intelligent design must complete these five tasks:

1. *Lay out reliable criteria for detecting intelligence.* It is not enough merely to point out that materialist explanations have thus far failed to account for certain types of biological complexity. To infer intelligence in such cases, it is necessary to have reliable criteria for detecting design. Intelligent causes can do things that undirected material causes cannot. Many special sciences (such as archeology, forensics and the search for extraterrestrial intelligence) already employ criteria to draw that distinction. The main criterion that intelligent design employs to detect design is *specified complexity.*

2. *Apply these criteria to biological systems.* Once criteria are in place

for detecting design, they need to be applied to actual biological systems. For instance, biochemist Michael Behe (*Darwin's Black Box,* 1996) connects specified complexity to biological design as follows. Behe defines a system as *irreducibly complex* if it consists of several interrelated parts where removing even one part destroys the system's function. For Behe, irreducible complexity is a sure indicator of design. One irreducibly complex biochemical system that Behe considers is the bacterial flagellum. I argue in *No Free Lunch* that irreducible complexity is a special case of specified complexity and that systems like the bacterial flagellum exhibit specified complexity and therefore are designed.

3. Show that these criteria effectively rule out material mechanisms. Criteria like specified and irreducible complexity purport to detect design. Yet if material mechanisms can plausibly give rise to systems that exhibit specified or irreducible complexity, there is no reason to attribute design. In biology, design becomes plausible only if material mechanisms can be effectively ruled out. To rule out material mechanisms is to say that material mechanisms are unable to produce an object in question. But what does it mean to attribute such an inability?

Behe, for instance, has argued that irreducibly complex biochemical machines, by being composed of numerous parts each of which is necessary for the system's function, could not be produced by gradual Darwinian means. To say that such systems "could not be produced" is to attribute an inability to the Darwinian mechanism. Yet evolutionary biologists are able to imagine scenarios in which such systems gradually incorporate parts that, though not originally necessary, become necessary over time.

Have evolutionary biologists thereby realistically assessed the Darwinian selection mechanism's ability, as actually operating in nature, to produce irreducible complexity? Or have they merely demonstrated the ability of their imaginations to conjure up how a Darwinian process might lead to such systems? Many evolutionary biologists are satisfied with a very undemanding form of ability or capacity—namely, *conceivability.* So long as they can *conceive* of a Darwinian or other material pathway to irreducible complexity, material mechanisms trump design.

Behe and the intelligent design community, by contrast, require a much more demanding form of ability or capacity in assessing whether the Darwinian mechanism, and material mechanisms generally, can produce irreducible complexity. For Behe, it's a probabilistic form in which highly improbable, functionally specified structures cannot happen by chance. This weds Behe's work on irreducible complexity to mine on specified complexity. The logical force of our argument purports to be the same as "You can't walk into a Las Vegas casino and get a hundred double-zeros in a row playing roulette." There's a sheer possibility that this could happen by chance but not a realistic one.

4. *Reconceptualize evolutionary biology within a design-theoretic framework.* In reconceptualizing evolutionary biology, intelligent design's main task is teasing apart the effects of intelligence from material mechanisms. Theodosius Dobzhansky famously remarked that nothing in biology makes sense apart from evolution. Rather than reject this claim outright, intelligent design digs deeper and asks whether evolution itself makes sense apart from the discernible influence of a designing intelligence. Material mechanisms are perfectly capable of explaining many small-scale evolutionary changes (such as insects developing insecticide resistance by means of natural selection). But where do material mechanisms break off and give way to intelligence? A mechanistic biology sees this as an illegitimate question. But a biology open to design is free to consider a broader set of questions, one that gives material mechanisms their full due but also allows the possibility of intelligent agency.

Design is always a matter of tradeoffs. Intelligent design helps us understand these tradeoffs and clarify the design problems that organisms actually face. This in turn keeps us from sweeping problems under the rug simply because evolution is purported to be a blind and wasteful process. A nonteleological approach to evolution has consistently led biologists to underestimate organisms. Is, for instance, junk DNA really junk? Work by John Bodnar and his associates suggests that some of it is not. Intelligent design continues to look for function where nonteleological approaches to evolution attribute clumsiness or incompetence.

Because intelligent design adds rather than removes tools from the biologist's tool chest (supplementing material mechanisms with intelligent agency), intelligent design can subsume present biological research. Even efforts to overturn the various criteria for detecting design are welcome within the intelligent design research program. (That's part of keeping the program honest.) Intelligent design can also function as a heuristic for guiding research, inspiring biologists to look for engineering solutions to biological problems that might otherwise escape them.

5. *Inspire a fruitful program of biological research that is uniquely design-theoretic.* The ultimate challenge facing intelligent design is to inspire a fruitful program of biological research that generates exciting new ideas and experiments for biology—ideas and experiments that would not have been possible assuming a purely materialist biology. Ideally, someone should be able to win a Nobel Prize for doing research that is specific to intelligent design and inconceivable apart from it.

One way this might happen is for organisms to exhibit designs that have no functional significance but that nonetheless give biological investigators insight into functional aspects of organisms. Such second-order designs would serve essentially as an "operating manual," of no use to the organism as such but of use to scientists investigating the organism. For now this remains a speculative possibility, but there are preliminary results from the bioinformatics literature that bear it out in relation to the protein-folding problem. (Such second-order designs appear to reside in the genome.) I don't mean to suggest that this is the only way intelligent design can become a research program that is simultaneously fruitful and design-specific. But while it makes perfect sense for a designer to throw in an "operating manual" (much as automobile manufacturers include operating manuals with the cars they make), this possibility makes no sense for blind material mechanisms, which cannot anticipate scientific investigators.

Where are we with these five tasks facing intelligent design? At present most of the discussion centers on tasks (1), (2) and (3). Most of the actual controversy is currently focused on the logical coherence of design-detecting criteria, their applicability to biology and their ability decisively to preclude material mechanisms. My own sense is that within the next five years design theorists will have identified and done a sufficiently thorough analysis of certain biological systems to show convincingly that these systems lie beyond the reach of material mechanisms. For now, however, design is back on the table as a live possibility confronting biology (despite the earnest hope of many biologists that design will simply vanish away). As for (4) and (5), (4) is just starting and (5) remains a promissory note. Intelligent design, like evolutionary biology, therefore has its own promissory notes. The difference is that intelligent design is only now being tried whereas evolutionary biology has been thoroughly tried and found wanting.

Ultimately, the success of intelligent design as a scientific project depends on making good on (5). Will that happen? I believe it will, but I don't have a sense of inevitability here. Design is, in my view, an extremely fruitful idea for biology and the natural sciences generally. But whether it bears fruit will depend on there being researchers to harvest the fruit. Given the intense pressures of a materialistic scientific establishment to overthrow intelligent design as well as the intense cultural pressures to use intelligent design as a weapon to overthrow materialism, the actual scientific program stands in danger of getting lost in the shuffle. I helped found the International Society for Complexity, Information and Design (<www.iscid.org>) to explore intelligent design's scientific merits. I'm heartened by progress to date, but I still await a series of breakthroughs that render intelligent design the clear winner in the scientific debate over biological origins.

38

MECHANISM

Since intelligent design is not a mechanistic theory of life's origin and development, how can it be scientific?

EVOLUTIONARY BIOLOGY TEACHES THAT ALL biological complexity is the result of material mechanisms. These include principally the Darwinian mechanism of natural selection and random variation but also include other mechanisms (symbiogenesis, gene transfer, genetic drift, the action of regulatory genes in development, self-organizational processes, etc.). These mechanisms are just that—mindless material mechanisms that do what they do irrespective of intelligence. To be sure, mechanisms can be programmed by an intelligence. But any such intelligent programming of evolutionary mechanisms is not properly part of evolutionary biology.

Intelligent design, by contrast, teaches that biological complexity is not exclusively the result of material mechanisms but also requires intelligence, where the intelligence in question is not reducible to such mechanisms. The central issue, therefore, is not the relatedness of all organisms, or what typically is called *common descent*. Indeed, intelligent design is perfectly compatible with common descent. Rather, the central issue is how biological complexity emerged and whether intelligence played an indispensable (which is not to say exclusive) role in its emergence.

Suppose, therefore, for the sake of argument that intelligence—one irreducible to material mechanisms—actually did play a decisive role in the emergence of life's complexity and diversity. How could we know it? Certainly specified complexity will be required. Indeed, if specified complexity is absent or in doubt, then the burden of evidence is on those who want to deny that material mechanisms can explain the object of inquiry. Only as specified complexity becomes assertible (see chapter fourteen) does the burden of evidence shift to those who want to continue to maintain that material mechanisms provide an adequate explanation.

In the face of this perfectly reasonable divvying of evidential burdens, evolutionary biology teaches that within biology the burden of evidence forever remains on those who want to deny the adequacy of material mechanisms. In fact, evolutionary biologists boldly proclaim that design is and always will be superfluous for understanding biological complexity. The only way actually to demonstrate this, however, is to provide detailed, testable accounts of how material mechanisms might actually explain the various forms of biological complexity out there. Now, if for representative instances of biological complexity such accounts could readily be produced, intelligent design would drop out of scientific discussion. Occam's razor, by proscribing superfluous causes, would in that case finish off intelligent design quite nicely.

But that hasn't happened. Why not? The reason is that there are plenty of complex biological systems for which no biologist has a clue how they emerged. I'm not talking about handwaving just-so stories. Biologists have plenty of those. I'm talking about detailed, testable accounts of how such systems could have emerged. To see what's at stake, consider how biologists propose to explain the emergence of the bacterial flagellum, a molecular machine that has become the mascot of the intelligent design movement.

In public lectures Harvard biologist Howard Berg calls the bacterial flagellum "the most efficient machine in the universe." The flagellum is a nano-engineered bidirectional motor-driven propeller on the backs of certain bacteria. It spins at tens of thousands of revolutions per minute, can change direction in a quarter turn and propels a bacterium through its watery environment. According to evolutionary biology it had to emerge via some material mechanism(s). Fine, but how?

The usual story is that the flagellum is composed of parts that previously were targeted for different uses and that natural selection then co-opted to form a flagellum. This seems reasonable, until we try to fill in the details. The only well-documented examples that we have of successful co-optation come from human engineering. For instance, an electrical engineer might co-opt components from a microwave oven, a radio and a computer screen to form a working television. But in that case, we have an intelligent agent who knows all about electrical gadgets and about televisions in particular.

But natural selection doesn't know a thing about bacterial flagella. So how is natural selection going to take extant protein parts and co-opt them to form a flagellum? The problem is that natural selection can only select for preexisting function. It can, for instance, select for larger finch beaks when the available nuts are harder to open. Here the finch beak is already in place and natural selection merely enhances its present functionality. Natural selection might even adapt a preexisting structure to a new function; for example, it might start with finch beaks adapted to opening nuts and end with beaks adapted to eating insects.

But for co-optation to result in a structure like the bacterial flagellum, we are not talking about enhancing the function of an existing structure or reassigning an existing structure to a different function. Rather, we are talking about reassigning multiple structures previously targeted for different functions to a novel structure exhibiting a novel function. Even the simplest bacterial flagellum requires around forty proteins for its assembly and structure. All these proteins are necessary in the sense that lacking any of them, a working flagellum does not result.

The only way for natural selection to form such a structure by co-optation, then, is for natural selection gradually to enfold existing protein parts into evolving structures whose functions co-evolve with the structures. We might, for instance, imagine a five-part mousetrap consisting of a platform, spring, hammer, holding bar and catch evolving as follows: it starts as a doorstop (thus consisting merely of the platform), then evolves into a tie-clip (by attaching the spring and hammer to the platform) and finally becomes a full mousetrap (by also including the holding bar and catch).

Design critic Kenneth Miller finds such scenarios not only completely plausible but also deeply relevant to biology. (In fact, he regularly sports a modified mousetrap cum tie-clip.) Intelligent design proponents, by contrast, regard such scenarios as rubbish. Here's why. First, in such scenarios the hand of human design and intention meddles everywhere. Evolutionary biologists assure us that eventually they will discover just how the evolutionary process can take the right and needed steps without the meddling hand of design. All such assurances, however, presuppose that intelligence is dispensable in explaining biological complexity. Yet the only evidence we have of successful co-optation comes from engineering and confirms that intelligence is indispensable in explaining complex structures like the mousetrap and, by implication, the flagellum. Intelligence is known to have the causal power to produce such structures. We're still waiting for the promised material mechanisms.

The other reason design theorists are less than impressed with co-optation concerns an inherent limitation of the Darwinian mechanism. The whole point of the Darwinian selection mechanism is that one can get from anywhere in biological configuration space to anywhere else, provided one can take small steps. How small? Small enough that they are reasonably probable. But what guarantee is there that a sequence of baby steps connects any two points in configuration space?

The problem is not simply one of connectivity. For the Darwinian selection mechanism to connect point A to point B in configuration space, it is not enough that there merely exist a sequence of baby steps connecting the two. In addition, each baby step needs to be in some sense "successful." In biological terms, each step requires an increase in fitness as measured in terms of survival and reproduction. Natural selection, after all, is the motive force behind each baby step, and selection only selects what is advantageous to the organism. Thus, for the Darwinian mechanism to connect two organisms, there must be a sequence of successful baby steps connecting the two.

Richard Dawkins compares the emergence of biological complexity to climbing a mountain—Mount Improbable, as he calls it (see his book *Climbing Mount Improbable*). He calls it Mount Improbable because if you had to get all the way to the top in one fell swoop (i.e., achieve a massive increase in biological complexity all at once), it would be highly improbable. But Mount Improbable does not have to be scaled in one leap. Darwinism purports to show how Mount Improbable can be scaled in small, incremental steps. Thus, according to Dawkins, Mount Improbable always has a gradual serpentine path leading to the top that can be traversed in baby steps. But such a claim requires verification. It might be a fact about nature that Mount Improbable is sheer on all sides and getting to the top from the bottom via baby steps is effectively impossible. A gap like that would reside in nature itself and not in our knowledge of nature. (It would not, in other words, constitute a god-of-the-gaps.)

Consequently, it is not enough merely to presuppose that a fitness-increasing sequence of baby steps connects two biological systems. It must be demonstrated. For instance, it is not enough to point out that some genes for the bacterial flagellum are the same as those for a type III secretory system (a type of pump) and then handwave that one was co-opted from the other. Anybody can arrange complex systems in series based on some criterion of similarity. But such series do nothing to establish whether the end evolved in a Darwinian fashion from the beginning unless each step in the series can be specified, the probability of each step can be quantified, the probability at each step turns out to be reasonably large and each step constitutes an advantage to the organism. (In particular, viability of the whole organism must at all times be preserved.) Only then do we have a mechanistic explanation (in Darwinian terms) of how one system arose from another.

Convinced that the Darwinian mechanism must be capable of doing such evolutionary design work, evolutionary biologists rarely ask whether such a sequence of successful baby steps in fact exists, much less attempt to quantify the probabilities involved. I attempt that in my book *No Free Lunch* (chapter five). There I lay out techniques for assessing the probabilistic hurdles that the Darwinian mechanism faces in trying to account for complex biological structures like the bacterial flagellum. The probabilities I calculate—and I try to be conservative—are horrendous and render natural selection utterly implausible as a mechanism for generating the flagellum and structures like it.

To sum up, evolutionary biology contends that material mechanisms are capable of accounting for all of biological complexity. Yet for biological systems that exhibit specified complexity, these mechanisms provide no explanation of how they were produced. Moreover, in contexts where the causal history is independently verifiable, specified complexity is reliably correlated with intelligence. At a minimum, biology should therefore allow the possibility of design in cases of biological specified complexity. But that's not how it is.

Evolutionary biology allows only one line of criticism against itself, namely, to show that a complex specified biological structure could not have evolved via any material mechanism. In other words, so long as some unknown material mechanism might (as a mere conceptual possibility) have evolved the structure in question, intelligent design is proscribed. This renders evolutionary theory immune to disconfirmation in principle because the universe of unknown material mechanisms can never be exhausted. The evolutionist accepts no burden of evidence. Instead of empirical evidence deciding between evolutionary biology and intelligent design, evolutionary biology is declared the winner by default. This isn't science. This is dogmatism.

39

TESTABILITY

Is intelligent design testable?
Is Darwinism testable?

NOTHING PREVENTS INTELLIGENT DESIGN OR DARWINISM from being testable. Imagine a world in which every bacterium spontaneously arranges its cell wall to display a novel chess position in which white delivers checkmate in four moves or less. (Suppose the standard Staunton chess pieces are clearly visible under an electron microscope.) Now imagine a world in which the fossil record showed perfect gradual progressions and in which natural selection and random variation could be observed in the laboratory to produce novel complex biological structures. Either of these scenarios would constitute a decisive joint test for intelligent design and Darwinism, the first positively confirming intelligent design, the second Darwinism.

The real world, however, is not nearly as clear-cut in favoring either intelligent design or Darwinism. In the real world, testability assumes a pragmatic rather than an in-principle or logical form. At the heart of testability is the idea that our scientific theories must make contact with and be sensitive to what's happening in nature. What's happening in nature must be able to affect our scientific theories not only in form and content but also in how much credence we attach to or withhold from them. For a theory to be immune to evidence from nature is a sure sign that we are not dealing with a scientific theory.

What then are we to make of the testability of intelligent design and Darwinism taken not in some unrealistic in-principle or mere-conceptual-possibility sense but practically and concretely? What are the specific tests for intelligent design? What are the specific tests for Darwinism? And how do the two theories compare in terms of testability? To answer these questions, let's run through several aspects of testability. These will include refutability, confirmation, predictability and explanatory power.

Let's begin with *refutability*. Refutability is my variant of Karl Popper's falsifiability. For Popper, a theory was testable if empirical data could in principle falsify it. The problem with falsifiability is that no scientific theory is, strictly speaking, falsifiable (unless of course it constitutes a logical contradiction, in which case its falsehood is independent of empirical considerations). In practice one can always shore up a failing scientific theory by adding suitably chosen auxiliary hypotheses that harmonize recalcitrant data with the theory. Sometimes adding auxiliary hypotheses is perfectly valid, as when one restricts the scope of a theory so that it applies only to relevant phenomena. (Consider the limitation of Newtonian mechanics to medium-sized objects at medium speeds. The theory works just fine for that range of phenomena, and neither quantum mechanics nor relativity theory improves the accuracy of Newtonian mechanics within that range.) On the other hand, adding auxiliary hypotheses can get highly artificial, as with Ptolemaic epicycles, where the presumed orbits of planets became increasingly complex to accommodate observations.

What about the falsifiability of intelligent design? If complex biological systems like the bacterial flagellum could be explicitly shown to result from material mechanisms (like the Darwinian mechanism), it would not follow as a logical entailment that the intelligent design of life is false in the sense of being necessarily untrue. One could, for instance, argue that a designer had designed the laws of physics and chemistry so that life would emerge by means of the Darwinian mechanism. In that case intelligent design would not, strictly speaking, be falsified. Biologists, however, would rightly discard it as superfluous.

The main point of Popper's criterion of falsifiability is not so much that scientific claims must have the possibility of being demonstrably false as that they must have the possibility of being eliminated as the result of new evidence. To underscore this point Popper even wrote a book titled *Conjectures and Refutations*. That is the point of refutability. If Behe's irreducibly complex biochemical machines suddenly submit to material mechanisms, design would become superfluous and drop out of scientific discussion. It would thus be refuted. Note that etymologically, *refutation* has nothing to do with *truth* or *falsehood*. The root idea of refutation is that of striking or beating down (as in "rebutting" or "butting heads"). Theories are refuted when, for whatever reason, they get beaten down and rejected—not because they are demonstrably false.

Refutability comes in degrees. Theories become more refutable to the degree that new evidence could render them unacceptable. Note that refutability asks to what degree theories could be refuted, not to what degree they actually have been refuted. Thus refutability gauges how sensitive theories are to refutation in principle rather than on the basis of any particular evidence. The more sensitive to evidence generally, the more refutable the theory. According to Popper, one mark of a good scientific theory is that it is highly refutable in principle while consistently unrefuted by the evidence in practice. Better yet are those theories on which scientists have expended tremendous diligence to refute them, only to have their efforts come to nothing. Within Popper's scheme of scientific rationality, theories are corroborated to the degree that they resist refutation.

Let's now ask, Is intelligent design refutable? Is Darwinism refutable? Yes to the first question, no to the second. Intelligent design could in principle be readily refuted. Specified complexity in general and irreducible complexity in biology are, within the theory of intelligent design, key markers of intelligent agency. If it could be shown that biological systems that are wonderfully complex, elegant and integrated—such as the bacterial flagellum—could have been formed by a gradual Darwinian process (and thus that their specified complexity is an illusion), then intelligent design would be refuted on the general grounds that one does not invoke intelligent causes when undirected natural causes will do. In that case Occam's razor would finish off intelligent design quite nicely.

By contrast, Darwinism seems effectively irrefutable. The problem is that Darwinists raise the standard for refutability too high. It is certainly possible to show that no Darwinian pathway could reasonably be expected to lead to an irreducibly complex biological structure like the bacterial flagellum. But Darwinists want something stronger, namely, to show that no conceivable Darwinian pathway could have led to that structure. Such a demonstration requires an exhaustive search of all conceptual possibilities and is effectively impossible to carry out. Yet the fact remains that for complex systems like the bacterial flagellum, no biologist has or is anywhere close to reconstructing its history in Darwinian terms (and not just its actual history but any detailed testable alternatives). Is Darwinian theory therefore refuted? Hardly. Darwinism is wonderfully adept at rationalizing its failures and therefore just keeps chugging along.

Refutability focuses on negative evidence against a theory—in this case, what it would take to bring down intelligent design or Darwinism. What about positive evidence for intelligent design and Darwinism? From the design theorist's perspective, the positive evidence for Darwinism is confined to small-scale evolutionary changes like insects developing insecticide resistance. This is not to deny large-scale evolutionary changes, but it is to deny that the Darwinian mechanism can account for them. Insects developing insecticide resistance confirms the Darwinian selection mechanism for small-scale changes but hardly warrants the grand extrapolation that Darwin's theory requires. It is a huge leap going from insects developing insecticide resistance via the Darwinian mechanism to their emergence by that same mechanism in the first place.

Darwinists shrug off the extrapolation from small-scale to large-scale evolution, arguing that it is a failure of imagination on the part of critics to appreciate the wonder-working power of the Darwinian mechanism when given vast time scales to operate. From the design theorist's perspective, however, this is not a case of failed imagination but of the emperor's new clothes. Yes, there is positive evidence for Darwinism, but the strength and relevance of that evidence on behalf of large-scale Darwinian evolution is very much under dispute (not just by design theorists but also by self-organizational theorists like Brian Goodwin, Stuart Kauffman and Mae-Wan Ho).

What about the positive evidence for intelligent design? As it turns out, biology is chock-full of specified complexity. And since specified complexity is a reliable empirical marker of actual design, that means biology is chock-full of evidence for design. The specified complexity exhibited in biological systems therefore provides positive evidence for intelligent design. Now it's true that specified complexity doesn't tell you what happened. It merely tells you that you have design, not how it got there. Nevertheless, the absence of a causal story for how design got there can't undermine a design inference. For instance, if you found a Buick Regal on Jupiter's moon Io, you would know that it was designed even if you had no idea how it got there. If SETI researchers ever discover a radio signal that decisively confirms an extraterrestrial intelligence, their knowledge of how the signal was produced would have no bearing on the validity of the design inference that they draw. If, for instance, a SETI researcher discovers a radio transmission of prime numbers from outer space, the inference to an extraterrestrial intelligence will be clear, but the researcher won't know "what happened" in the sense of knowing any details about the radio transmitter or, for that matter, any details about the extraterrestrial that built the radio transmitter and sent the radio transmission.

The usual counterargument here is to point out that we have experience with radio transmitters. At least with extraterrestrial intelligences we can guess what might have happened. But we do not have any experience with unevolved designers, and that is clearly what we would be dealing with in biology. Actually, if an unevolved designer were responsible for biological complexity, we would have quite a bit of experience with such a designer through the designed objects in nature that confront us most—namely, plants, animals and not least fellow human beings—provided, of course, such biological design is actual. Since the design of biological systems is precisely the question at issue, to argue that we have no experience observing the designs of an unembodied designer is mere question begging. On the other hand, it is true that we possess very little insight at this time into how such an unevolved designer acted to bring about the complex biological systems that have emerged over the course of natural history.

Darwinists take this present lack of insight into the workings of an unevolved designer not as remediable ignorance and not as evidence that the designer's capacities far outstrip ours but as proof that there is no unevolved designer, period. By the same token, if an extraterrestrial intelligence communicated via radio signals with earth and solved computational problems that exceeded anything an ordinary or quantum computer could ever solve, we would have to conclude that we were not really dealing with an intelligence because we have no experience of supermathematicians who can solve such problems.

With respect to biological design, humans are in the same position as is a dog sitting in the back of a pickup. The dog hasn't the foggiest idea how the pickup was put together and doesn't know any other dogs who understand how it was put together. Our incomprehension over biological design is the incomprehension of the dog in the pickup. We don't know how the designer chose to go about putting together biological systems, and if we're scientific naturalists, we will assiduously avoid facing their specified complexity. It's regarding this last point that the dog analogy breaks down. The dog, given its limited cognitive apparatus, can't help but miss the specified complexity of the pickup. On the other hand, humans, though not privy to the construction history of biological systems, have to will ignorance of their specified complexity. This combination of unwilled and willed ignorance of the subtlety of biological design renders Darwinism plausible and enables naturalists to reject design without compunction. Ironically, Darwinists regularly sing the praises of natural selection and the wonders it has wrought while admitting that they have no comprehension of how it wrought those wonders. Natural selection, we are assured, is far cleverer than we are or can ever hope to be. Darwinists have merely exchanged one form of awe for another. They have not eliminated it.

In short, it is no objection that at this time we do not comprehend how an unevolved designer produced biological systems exhibiting specified complexity any more than we had to withhold a design inference with regard to the pyramids before archeologists posited a reasonable explanation of how the enormous structures were put together. We know that specified complexity is reliably correlated with the effects of intelligence. The only reason to insist on looking for nontelic explanations to explain the complex specified structures of biology is a prior commitment to naturalism that perforce excludes unevolved designers. It is utterly bogus to claim on a priori grounds that such entities do not exist or to claim that if they do exist, they can have no conceivable relevance to what happens in the world. Such a precommitment is gratuitous to the practice of science. Specified complexity confirms design regardless of whether the designer responsible for it is evolved or unevolved.

Another aspect of testability is predictability. A good scientific theory, we are told, is one that predicts things. If it predicts things that do not happen, then it is tested and found wanting. If it predicts things that do happen, then it is tested and regarded as successful. If it does not predict anything, however, what then? Often with theories that try to account for features of natural history, prediction gets generalized to include retrodiction, in which a theory also specifies what the past should look like. Darwinism is said to apply retrodictively to the fossil record and predictively in experiments that place organisms under selection pressures to induce adaptive changes.

But in fact Darwinism does not retrodict the fossil record, much less predict adaptive complexity-increasing change. That's not to say Darwinism specifies a particular pattern of the fossil record or a particular pattern of adaptive complexity-increasing change and gets it wrong. Rather, it doesn't specify either of these at all (whether correctly or incorrectly). For instance, natural selection and random variation applied to single-celled organisms offers no insight at all into whether we can expect multicelled organisms, much less whether evolution will produce the various body-plans of which natural history has left us a record. At best one can hope that there is consilience, namely, that the broad sweep of evolutionary history as displayed in the fossil record is consistent with Darwinian evolution. (See, for instance, E. O. Wilson's *Consilience: The Unity of Knowledge*.)

But design theorists strongly dispute Darwinism's supposed consilience as well, pointing especially to the Cambrian explosion, which remains a complete mystery from the vantage of Darwinism. Detailed retrodiction and detailed prediction are not virtues of Darwin's theory. Organisms placed under selection pressures either adapt or go extinct. Except in the simplest cases where there is, say, some point mutation that reliably confers antibiotic resistance on a bacterium, Darwin's theory has no way of predicting just what sorts of adaptive changes will occur. "Adapt or go extinct" is not a prediction of Darwin's theory but a logical truth that can be reasoned out independently of the theory. Indeed, it was reasoned out before Darwin and reasoned out by design advocates no less.

Darwin's theory has virtually no predictive power. Insofar as it offers predictions, they are extremely general, concerning the broad sweep of natural history, and even in that respect they are highly questionable. Indeed, why else did Stephen Jay Gould and Niles Eldredge need to introduce their theory of punctuated equilibrium if the fossil record was such an overwhelming vindication of Darwinism? On the other hand, when Darwinism's predictions are not extremely general, they are extremely specific and picayune, dealing with small-scale adaptive changes that in no way demonstrate macroevolutionary change but instead revisit facts of nature observed and accepted before Darwin. For instance, in the 1830s Edward Blyth wrote two articles in which he described the process by which nature keeps organisms within certain bounds and maintains quality control. Blyth was, of course, describing natural selection, though without, like Darwin, divinizing it into the great creative force responsible for living forms. Newton was able to predict the precise pathways that planets trace out in cosmic history. Darwinists can neither predict nor retrodict the precise pathways that organisms trace out in the course of natural history.

But what about the predictive power of intelligent design? Intelligent design offers one obvious prediction, namely, that nature should be chock-full of specified complexity and therefore should contain numerous pointers to design. That prediction is increasingly confirmed. What's more, once designed systems are in place, operational and interacting (as within an economy or ecosystem), intelligent design predicts certain patterns of technological evolution, notable among these being sudden emergence, convergence to local optima and extinction. Although research in this area is only now beginning, preliminary indications are that biology confirms these patterns of technological evolution. Significantly, these patterns are non-Darwinian.

The work of Genrich Altshuller is seminal in this regard. Altshuller, a Russian engineer and scientist, analyzed over 400,000 patents from different fields of engineering. He was especially interested in tracking the evolution of technological systems. From the trends he observed, he found that the evolution of engineering systems is not random but obeys certain laws. He codified these laws under the acronym TRIZ. TRIZ corresponds to a Russian phrase that in English means "theory of inventive problem solving" (and is sometimes given the acronym TIPS). Even though TRIZ is widely employed in industry, its applications to biology are only now becoming evident.

Even so, there is a sense in which to require prediction of intelligent design fundamentally misconstrues intelligent agency and design. To require of intelligent design that it predict specific novel instances of design in nature is to put design in the same boat as natural laws, locating their explanatory power in an extrapolation from past experience. This is to commit a category mistake. To be sure, designers, like natural laws, can behave predictably. (Designers often obey policies and conventions and follow routines in problem solving.) Yet unlike natural laws, which are universal and uniform, designers are also innovators. Specific innovations can't be predicted. At best what can be predicted are certain trends, as with, for instance, Intel and their all-but-guaranteed invention of faster and faster computer chips. But what we are actually dealing with here is the outworking and improvement of an existing technology, not its actual invention. Designers are inventors. We cannot predict what an inventor will invent, short of becoming that inventor.

Intelligent design is therefore radically at odds with a mechanistic science wedded solely to undirected natural causes ruled by unbroken natural laws. Such laws don't merely permit but in fact require predictability of the phenomena within their purview. Intelligent causes, by contrast, though often willing to go along with certain rules or laws, can never in any ultimate sense be bound by them. By taking intelligent causes seriously, intelligent design therefore challenges the undue emphasis that a mechanistic science places on predictability. To be sure, intelligent design rightly predicts the presence of design in nature and the technological evolution of existing designs. But intelligent design does not predict, cannot predict and has no obligation to predict the particulars of fundamentally new designs.

Finally, let us turn to the last aspect of testability—explanatory power. According to Darwin, the great advantage of his theory over William Paley's theory of design was that Darwin's theory managed to account for a wide diversity of biological facts that Paley's theory could not. Darwin's theory was thus thought to have greater explanatory power than Paley's,

and this relative advantage could be viewed as a joint test of the two theories. Underlying explanatory power is a view of explanation known as
inference to the best explanation, in which a "best explanation" always
presupposes at least two competing explanations. Consequently, a "best
explanation" is one that comes out on top in a competition with other explanations. Design theorists see advances in the biological and information sciences as putting design back in the saddle and enabling it to
out-explain Darwinism, thus making design rather than natural selection
currently the best explanation of biological complexity. Darwinists, of
course, see the matter differently.

What I want to focus on here, however, is not the testing of Darwinism
and intelligent design against the broad body of biological data, but the related question of which theory can in principle accommodate the greater
range of biological possibilities. Darwinism and intelligent design are not
just theories that make claims about the world (claims that can be either
true or false, assertible or unassertible). They are also theoretical tool
chests offering various explanatory tools and strategies. Darwinism's tool
chest is thoroughly naturalistic and allows for a certain set of explanatory
options. Intelligent design's tool chest is not constrained by naturalism
and allows for a wider set of explanatory options. Are there things that
might occur in biology for which a design-theoretic tool chest could provide a better, more accurate explanation than a purely Darwinian and
therefore nonteleological tool chest? The answer is yes.

First off, let us be clear that intelligent design, conceived now not as a
theory but as a tool chest, can accommodate all the results of Darwinism.
To be sure, as scientific theories, Darwinism and intelligent design contradict each other since intelligent design claims biology exhibits actual design whereas Darwinism claims biology exhibits only apparent design.
But as a tool chest, intelligent design incorporates all the tools of Darwinism. Intelligent design assigns a high place to natural causes and mechanisms. Insofar as these operate in nature, intelligent design wants to understand them and give them their due. But intelligent design also regards
natural causes as incomplete and leaves the door open to intelligent
causes. Intelligent design therefore does not repudiate the Darwinian
mechanism. It merely assigns it a lower status.

The Darwinian mechanism does operate in nature, and, insofar as it
does, intelligent design can accept what it delivers. Even if the Darwinian
mechanism could be shown to do all the design work for which design
theorists want to invoke intelligent causation (say for the bacterial flagel-

lum and systems like it), a design-theoretic tool chest would not obviate any valid findings of science. To be sure, much as some tools just sit there never to be used, design would then become superfluous. But a design-theoretic tool chest would not on this account collapse on itself through internal contradiction.

The worst that can happen to a design-theoretic tool chest is that design becomes a superfluous component of it. The worst that can happen to a Darwinian tool chest is that scientists, by limiting themselves to it, miss the design that actually is present in nature and thus fundamentally misconstrue reality. The dangers that confront science by adopting a Darwinian tool chest and the naturalism that motivates its choice of tools therefore far outweigh the dangers that confront science by adopting a design-theoretic tool chest.

To see this, suppose that I were a supergenius molecular biologist and that I invented some hitherto unknown molecular machine that is far more complicated and marvelous than the bacterial flagellum. Suppose, further, I inserted the genes for this machine into a bacterium, set this genetically modified organism free and allowed it to reproduce in the wild. Finally, suppose I destroyed all evidence of my having created the molecular machine. Suppose, for instance, the machine is a nano-engineered syringe that injects other bacteria and sucks out their contents for food. (I am not familiar with any such molecular machine in the wild.)

Question: if a Darwinist came upon this bacterium with its novel molecular machine in the wild, would that machine be attributed to design or to natural selection? When I presented an example like this to David Sloan Wilson at a conference at MIT in 1999, he shrugged it off and remarked that natural selection created us and so by extension also created my novel molecular machine. But of course this argument will not wash since the issue is whether natural selection could indeed create us. Wilson begged the question. What's more, if Darwinists came upon my bacterial syringe in the wild—composed of proteins and coded for by DNA in exactly the same way as other, naturally occurring biochemical machines—they would not look to design but would reflexively turn to natural selection.

But, if we go with the story, I designed the bacterial syringe, and natural selection played no role in its formation. Moreover, intelligent design, by focusing on the syringe's specified complexity, would confirm the syringe's design whereas Darwinism never could. It follows that a design-theoretic tool chest could bring to light biological facts that would forever remain invisible given a strictly Darwinian tool chest. This possibility constitutes a joint test of Darwinism and intelligent design that strongly supports intelligent design—if not as the truth then certainly as a live theoretical option that must not be precluded for a priori philosophical reasons like naturalism. In fact, this example points up that there are good practical reasons to take intelligent design seriously. We are on the cusp of a bioengineering revolution whose fallout is likely to include bioterrorism. Thus we can expect to see bioterrorism forensics emerge as a practical scientific discipline. How will such forensic experts distinguish the terrorists' biological designs from other biological designs?

To sum up, there is no merit to the charge that intelligent design is untestable or has not put forward any "testable models." Intelligent design's claims about specified and irreducible complexity are in close contact with the data of biology and open to refutation as well as confirmation. What's more, as a tool chest for scientific inquiry, intelligent design is more robust and sensitive to the possibilities that nature might actually throw our way than Darwinism, which must view everything through the lens of chance and necessity and take a reductive approach to all signs of teleology in nature.

40

THE SIGNIFICANCE
OF MICHAEL BEHE

*Why do evolutionary biologists think
Michael Behe's work on irreducible
complexity has been discredited?*

GET ON THE INTERNET AND READ SOME of the criticisms of Michael Behe's work, and you'll think he is a crank, a fraud and a knave. His work, we are told, has been "thoroughly discredited," "completely demolished" and "utterly destroyed." Critics reluctantly concede that his discussions of biochemistry (Behe's field of expertise) are unobjectionable. But when it comes to his definition and use of irreducible complexity, critics regard him as misguided and safely to be ignored. Critics contend that the biological community has carefully considered Behe's work, found it deeply flawed and therefore rejected it. But in fact the biological community is still coming to terms with Behe's work. Behe has focused attention on a major conceptual problem in evolutionary biology. The problem was noted before, but not in so stark a form.

Behe's challenge has been so unsettling that many in the biological community find it easier to pretend his work has been discredited than actually to engage it. The biology department at one well-regarded evangelical Christian institution is a surprising case in point. Its biology faculty, by last report, remain adamantly opposed to Behe and intelligent design but at the same time have explicitly refused even to read his work lest they dignify it with their time and attention. And so a convenient fiction has emerged in which biologists continually reassure each other that Behe has been refuted but either fail to provide an actual refutation or attack a caricature of Behe's case against Darwinian evolution. Yet to a dispassionate outsider, it's clear that something significant is afoot. If the worst humiliation is not to be taken seriously, then Behe is being taken all too seriously.

Indeed, Behe has attracted a band of vocal and passionate critics who engage him at length. The controversy centers on a book that Behe published in 1996, *Darwin's Black Box*. This widely influential book opened a great many ideas, central among them the concept of irreducible complexity. As Behe defines it, an integrated multipart functional system is irreducibly complex if removing any of its parts destroys the system's function. Critics have interpreted Behe's use of this concept in one of two ways, neither of which does justice to Behe's project. Thus critics see Behe as making either a purely logical or a purely empirical point. The logical point is this: Certain structures are provably inaccessible to a Darwinian mechanism. They have property P (i.e., irreducible complexity). But certain biological structures also have property P, so they, too, must be inaccessible to a Darwinian mechanism. The empirical point is this: Certain biological structures are awfully complicated. There is not even a suggestion in the literature concerning how the Darwinian mechanism might construct them. So chances are that something beyond natural selection was responsible for their creation.

So stated, these are fundamentally different points and involve very different questions. If Behe seeks to make a purely logical point, then his model needs to be rigorous and mathematical after the fashion of Noam Chomsky's demonstration that, for example, finite state automata are incapable of generating certain languages. If he wishes to make a purely empirical point, then he wastes his time bringing in the notion of irreducible complexity when what he really means is simply that the evolutionary pathways of certain biological objects have yet to be adequately explained. According to critics, the conflation of these two different theses, the logical and the empirical, works rhetorically, but for a bad reason: it suggests in virtue of the sonority of the words *irreducible complexity* that something rigorous or well-defined is at issue when what is really at issue, provided Behe has abjured the logical point, is what has always been at issue between Darwinists and their critics—the idea that life is simply too complicated to result from a blind, undirected, hit-or-miss, trial-and-error Darwinian process.

According to Darwinists, neither the logical point nor the empirical point nor a conflation of the two poses any challenge to their theory. Let's consider these options in turn. As for the logical point, irreducible complexity clearly cannot close off all logically possible avenues of Darwinian evolution. What irreducible complexity says is that all parts of a system are indispensable in the sense that if you remove a part and don't alter the other parts, you cannot recover the original function of the system. But that leaves the possibility of removing parts and modifying others to recover the original function. Also it leaves the possibility of removing parts and isolating subsystems that serve some other function (a function that could conceivably be subject to selection pressure). Irreducible complexity, treated as a logical restriction, therefore leaves some loopholes for the Darwinian mechanism. (Critics sometimes portray Behe as denying this point, but in fact Behe never denied such logically possible loopholes.)

As for the empirical point, it seems merely to commit the standard fallacy of arguing from ignorance. So what if certain biological systems are incredibly complicated and we haven't figured out how they originated? That doesn't mean the Darwinian mechanism or some other material mechanism didn't do it. It may just mean that we haven't figured out how those mechanisms did it quite yet. And as for conflating the logical and empirical points, that's the most disreputable option of all, for it makes Behe and fellow design theorists guilty of equivocation, of using irreducible complexity to make a logical or empirical point as expedience dictates.

But this is too easy. In fact, Behe's project is more subtle than any of these criticisms suggests. Behe's project is properly conceived as making three key points: a logical, an empirical and an explanatory point. What's more, he conflates none of them. The logical point is this: Certain artificial structures are provably inaccessible to a *direct* Darwinian pathway because they have property P (i.e., irreducible complexity). But certain biological structures also have property P, so they, too, must be inaccessible to a direct Darwinian pathway. This formulation looks similar to the previous logical point, but it differs in one crucial respect. In the previous formulation, inaccessibility was with respect to the Darwinian mechanism in toto and therefore with respect to all Darwinian pathways whatsoever, both direct and indirect. Here, the restriction is only on direct Darwinian pathways.

A direct Darwinian pathway is one in which a system evolves by natural selection, incrementally enhancing a given function. As the system evolves, the function does not. Thus we might imagine that in the evolution of the heart, its function from the start was to pump blood. In that case a direct Darwinian pathway might account for it. On the other hand, we might imagine that in the evolution of the heart its function was initially to make loud thumping sounds to ward off predators and that only later did it take on the function of pumping blood. In that case an indirect Darwinian pathway would be needed to account for it. Here the pathway is indirect because not only does the system evolve but so does the system's function. Now, as a logical point, Behe was only concerned with direct Darwinian pathways. This becomes immediately evident from reading *Darwin's Black Box* since in his definition of irreducible complexity, the function of the system in question always stays put.

Does Behe's definition of irreducible complexity render certain structures provably inaccessible to direct Darwinian pathways? As laid out in *Darwin's Black Box*, Behe's definition actually needed a little fine-tuning. The problem is that Behe didn't address systems that could retain their function by removing parts and then modifying the other parts that remained. (Behe considered only removal, not modification.) But there's a quick fix here, which I describe in chapter five of *No Free Lunch*, and that is simply to strengthen the concept of irreducible complexity to include a minimal complexity condition. Essentially this condition says that the system cannot be simplified and still retain the level of function needed for selective advantage. With this proviso, irreducible complexity logically rules out direct Darwinian pathways. Note that many of the irreducibly complex systems Behe considers (notably the bacterial flagellum) satisfy this proviso.

In ruling out direct Darwinian pathways to irreducibly complex systems, Behe isn't saying it's logically impossible for the Darwinian mechanism to attain such systems. It's logically possible for just about anything to attain any other thing via a vastly improbable or fortuitous event. For instance, it's logically possible that with my very limited chess ability I might defeat the reigning world champion, Vladimir Kramnik, in ten straight games. But if I do so, it will be despite my limited chess ability and not because of it. Likewise, if the Darwinian mechanism is the means by which a direct Darwinian pathway leads to an irreducibly complex biochemical system, then it is despite the intrinsic properties or capacities of that mechanism. Thus, in saying that irreducibly complex biochemical systems are provably inaccessible to direct Darwinian pathways, design proponents are saying that the Darwinian mechanism has no intrinsic capacity for generating such systems except as vastly improbable or fortuitous events. Accordingly, to attribute irreducible complexity to a direct Darwinian pathway is like attributing Mount Rushmore to wind and erosion. There's a sheer possibility that wind and erosion could sculpt Mount Rushmore, but not a realistic one.

With direct Darwinian pathways ruled out, that leaves indirect Darwinian pathways. Here Behe's point is no longer logical but empirical. The fact is that for irreducibly complex biochemical systems, no indirect Darwinian pathways are known. At best, biologists have been able to isolate sub-systems of such systems that perform other functions. But any reasonably complicated machine always includes subsystems that perform functions distinct from the original machine. So the mere occurrence or identifica-tion of subsystems that could perform some function on their own is no evidence for an indirect Darwinian pathway leading to the system. What's needed is a seamless Darwinian account that's both detailed and testable of how subsystems undergoing coevolution could gradually transform into an irreducibly complex system. No such accounts are available or forthcoming. Indeed, if such accounts were available, critics would merely need to cite them, and intelligent design would be finished.

Critics of Behe are at this point quick to throw the argument-from-igno-rance objection his way, but this criticism can't be justified. A common way to formulate this criticism is to say, "Absence of evidence is not evidence of absence." But as with so many overused expressions, this one requires nuancing. Certainly this dictum appropriately characterizes many every-day circumstances. Imagine, for instance, someone feverishly hunting about the house for a missing set of car keys, searching under every object, casing the house, bringing in reinforcements and then, the next morning, when all hope is gone, finding them on top of the car outside. In this case the absence of evidence prior to finding the car keys was not evidence of absence. Yet with the car keys there was independent evidence of their ex-istence in the first place.

But what if we weren't sure that there even were any car keys? The sit-uation in evolutionary biology is even more extreme than that. One might not be sure our hypothetical set of car keys exist, but at least one has the reassurance that car keys exist generally. Indirect Darwinian pathways are more like the supposed leprechauns that Johnny is certain are hiding in his room. Imagine this child were so ardent and convincing that he set all of Scotland Yard, indeed some of the best minds of the age, onto the task of searching meticulously, tirelessly, decade after decade, for these supposed leprechauns, for any solid evidence at all of their prior habita-tion of the bedroom. And then imagine that in all those decades, the de-tectives, driven by gold fever for the leprechaun's treasure, let's say, never flagged in searching out and postulating new ways of catching a glimpse of a leprechaun, a leprechaun hair, a leprechaun fingerprint, any solid clue at all. After these many decades, with not a single solid clue to show for all that work, what should one say to the aging parents of the now ag-ing boy if these parents decided there were no leprechauns in the boy's room? Would it be logical to shake your finger at the parents and tell them, "Absence of evidence is not evidence of absence. Step aside and let the experts get back to work." That would be absurd. And yet that, essentially, is what evolutionary biologists are telling us concerning that utterly fruitless search for credible indirect Darwinian pathways to account for irreducible complexity.

If after repeated attempts you don't find what you expect to find after looking in all the right places and if you never had any evidence that the thing you were looking for existed in the first place, then you have reason to think that the thing you are looking for doesn't exist at all. That's precisely Behe's point about indirect Darwinian pathways. (See his chapter in *Darwin's Black Box* titled "Publish or Perish.") It's not just that we don't know of such a pathway for, say, the bacterial flagellum (the irreducibly complex biochemical machine that has become the mascot of the intelligent design movement). It's that we don't know of such pathways for any such systems. The absence here is pervasive and systemic. That's why critics of Darwinism like Franklin Harold and James Shapiro (neither of which is an intelligent design supporter) argue that positing as-yet-undiscovered indirect Darwinian pathways for such systems constitute "wishful speculations."

Behe's logical point is that irreducible complexity renders biological structures provably inaccessible to direct Darwinian pathways. Behe's empirical point is that the failure of evolutionary biology to discover indirect Darwinian pathways leading to irreducibly complex biological structures is pervasive and systemic, and that such a failure is reason to doubt that indirect Darwinian pathways are the answer to irreducible complexity. The logical and empirical points together constitute a devastating indictment of the Darwinian mechanism, which has routinely been touted as capable of solving all problems of biological complexity once an initial life form is on the scene. Even so, the logical and empirical points together don't answer how one gets from the failure of Darwinism to account for irreducibly complex systems to the legitimacy of employing design to account for them.

This is where the third main point of Behe's project—Behe's explanatory point—comes in. Scientific explanations come in many forms and guises, but the one thing they cannot afford to be without is causal adequacy. A scientific explanation needs to invoke causal powers sufficient to explain the effect in question. Otherwise, the effect is unexplained. The effect in question for Behe is the irreducible complexity of certain biochemical machines. How did such systems come about? Not by a direct Darwinian pathway, for irreducible complexity rules that out on logical grounds. And apparently not by indirect Darwinian pathways either, for the absence of scientific evidence here is complete. (Critics who claim otherwise are bluffing.) What's more, appealing to unknown material mechanisms is even more tenuous.

Thus, when it comes to irreducibly complex biochemical systems, there's no evidence that material mechanisms are causally adequate to bring them about. But what about intelligence? It is well known that intelligence produces irreducibly complex systems. (For example, humans regularly produce machines that exhibit irreducible complexity.) Intelligence is thus known to be causally adequate to bring about irreducible complexity. Behe's explanatory point, therefore, is that on the basis of causal adequacy, intelligent design is a better scientific explanation than Darwinism for the irreducible complexity of biochemical systems.

Behe's logical and empirical points are mainly negative: they focus on limitations of the Darwinian mechanism. Behe's explanatory point, by contrast, is positive: it provides positive grounds for thinking that irreducibly complex biochemical systems are in fact designed. One question about these points is now likely to remain. Behe uses the logical point to rule out direct Darwinian pathways and the empirical point to rule out indirect Darwinian pathways to irreducible complexity. But the absence of empirical evidence for direct Darwinian pathways leading to irreducible complexity is as complete as for indirect Darwinian pathways. It might seem, then, that the logical point is superfluous inasmuch as the empirical point dispenses with both types of Darwinian pathways. But in fact the logical point helps tighten the noose around Darwinism in a way that the empirical point can't.

If you look at the best confirmed examples of Darwinian evolution in the literature (from Darwin to the present), what you find is natural selection steadily improving a given feature that is performing a given function in a given way. Indeed, the very notion of "improvement" (which plays such an important role in Darwin's *Origin of Species*) typically connotes that a given thing is getting better in a given respect. Improvement in this sense corresponds to a direct Darwinian pathway. By contrast, an indirect Darwinian pathway (where one function gives way to another function and thus can no longer improve because it no longer exists), though often inferred by evolutionary biologists from fossil or molecular data, tends to be much more difficult to establish rigorously.

The reason is not hard to see: By definition natural selection selects for preexisting function. It cannot select for future function. Once a novel function is realized, the Darwinian mechanism can select for it as well. But making this transition is the hard part. How does one evolve from a system exhibiting a preexisting selectable function to a new system exhibiting a novel selectable function? Natural selection is no help here, and all the weight is on random variation to come up with the right and needed modifications during the crucial transition time when functions are changing (or, as Darwin put it in his *Origin of Species,* "unless profitable variations do occur, natural selection can do nothing"). The actual evidence that random variation can produce the successive modifications needed to evolve irreducible complexity is nil.

Behe's logical point about irreducible complexity ruling out direct Darwinian pathways therefore rules out the form of Darwinian evolution that is best confirmed. What's more, it rules out the only form of Darwinian evolution that is open to logical analysis. Indirect Darwinian pathways, by contrast, are so open-ended that no logical analysis is capable of constraining them. (Almost invariably they are left unspecified, thus rendering them neither falsifiable nor testable.) Behe's logical point therefore takes logic as far as it can in constraining the Darwinian mechanism and leaves empirical considerations to rule out what remains. And since logical inferences are inherently stronger than empirical inferences, Behe has made his critique of the Darwinian mechanism as strong and tight as possible. It's not just that certain biological systems are so complex that we can't imagine how they evolved by Darwinian pathways. Rather, we can show conclusively that they could not have evolved by direct Darwinian pathways and that indirect Darwinian pathways, which have always been on much less stable ground, are utterly without empirical support.

To sum up, Behe is significant in the debate between intelligent design and Darwinism because he has taught us how to evaluate the relative merits of each. He has done this by giving us the concept of irreducible complexity and by showing us how to employ it. By carefully analyzing and disentangling the logical, the empirical and the explanatory implications of irreducible complexity for the Darwinian mechanism, Behe has demonstrated that intelligent design is at the very least a viable contender in any attempt to explain the irreducible complexity of biochemical systems. What's more, he has shown how to bridge the scientific theory of design with our commonsense intuitions about design. In media reports on intelligent design, one often hears the following sound bite: "Life is too complicated to have arisen by natural forces, so it must have been designed." This sound bite captures many people's intuitions about intelligent design, but it is too simplistic for scientific purposes. Behe has shown us how to interpret this claim, substituting the rigorously defined phrase *irreducibly complex* for the vague and undefined phrase *too complicated,* and he has shown us how to reason our way properly from the inadequacy of undirected natural forces to design.

41

PEER REVIEW

*If intelligent design is a scientific research
program, why don't design theorists
publish or have their work cited in the peer-
reviewed literature?*

THE CLAIM THAT DESIGN THEORISTS DO NOT PUBLISH or have their work cited in the peer-reviewed literature is false. In fact, it is false any way one interprets that claim. The International Society for Complexity, Information and Design has emerged as the professional society of the intelligent design community. (See <www.iscid.org>.) That society, at the time of this writing, lists over fifty research fellows. The fellows of the society include full-fledged senior faculty at such schools as Oxford University in England, Princeton University in the United States, the University of New Brunswick in Canada, the University of Sydney in Australia, the University of Auckland in New Zealand, Hanyang University in Korea, Helsinki University of Technology in Finland and the State University of Applied Sciences in Frankfurt, Germany. (See <www.iscid.org/fellows.php>.)

The ISCID fellows cover the gamut of disciplines, including the full range of natural sciences. All the fellows are distinguished researchers in their own right and have published extensively in the peer-reviewed literature in their respective disciplines. Fritz Schaefer, the inventor of computational quantum chemistry, stands out. With over nine hundred peer-reviewed publications, he is the third-most-cited chemist in the world and has been considered for the Nobel Prize five times. Hence there is no question that design theorists publish and have their work cited in the peer-reviewed literature. They are credible scientists and scholars.

Of course, the real question is whether design theorists publish work that supports intelligent design in the peer-reviewed literature. Here again there is no problem. Readers may refer to the ISCID bibliography (at <www.iscid.org/bibliography/bibliography.php>) for a list of works in the peer-reviewed literature by design theorists who support intelligent design. (Note that this is a bibliography of design-relevant literature and thus also includes references to work by scientists who are not design theorists.) Rather than list a number of such works, it may be instructive for me here to describe the peer-review process for my book *The Design Inference* because it points up intelligent design's progress in breaking into the peer-reviewed literature as well as the obstacles we face.

The Design Inference appeared in Cambridge University Press's monograph series Cambridge Studies in Probability, Induction and Decision Theory. This series is the equivalent of a journal. It has a general editor, Brian Skyrms (who is a member of the National Academy of Sciences). It also has an editorial board, which at the time of publication consisted of the following: Ernest Adams, Ken Binmore, Jeremy Butterfield, Persi Diaconis, William Harper, John Harsanyi (who in 1994 shared the Nobel Prize in economics with John Nash, the protagonist of *A Beautiful Mind*), Richard Jeffrey, Wolfgang Spohn, Patrick Suppes, Amos Tversky and Sandy Zabell. This editorial board constitutes a literal who's who in the world of statistics and inductive reasoning.

The Design Inference went to three anonymous referees for a grueling yearlong review process. The first referee was overwhelmingly positive. The second referee was on balance negative, though he or she had some positive things to say about the manuscript. The general editor wanted the book in his series and therefore gave it to a third referee as a tiebreaker. The third referee was very positive about the manuscript but wanted significant revisions. (The referee report was seven single-spaced pages.) I agreed to do the revisions, whereupon the book was recommended to the Cambridge Syndicate, which in turn then issued me a contract to publish the monograph. The review/referee process for *The Design Inference* was more rigorous than anything I've experienced in the peer-reviewed journals in which I've published, and that includes math, philosophy and theology journals. The only reason *The Design Inference* didn't appear in a journal is that the argument required a book-length treatment. This book has been widely cited, including favorable citations in the peer-reviewed scientific literature (e.g., *The International Journal of Fuzzy Systems*).

Although peer review was a good thing for *The Design Inference*, I decided to forego peer review for its sequel, *No Free Lunch*. While I was still writing *No Free Lunch*, I contacted Cambridge University Press about publishing this book as a sequel to *The Design Inference*. Because *The Design Inference* had been Cambridge University Press's bestselling philosophical monograph in several years, it seemed likely that they would be interested in a follow-up volume. I wanted a contract for this book on the basis of a prospectus and some sample chapters, not an uncommon request for a sequel to a highly successful monograph. I sought this so that I wouldn't have to wait almost two and a half years between the time I submitted the completed manuscript and its publication, as in the case of *The Design Inference*. My work was being widely discussed, and I wanted the sequel to appear without delay.

The New York editor at Cambridge (not Skyrms) informed me that even though *The Design Inference* was one of their bestsellers, it was controversial; and even though the press didn't mind controversy as such, it had come to light that I was being labeled a "creationist." Thus, before Cambridge University Press could issue a contract, I would have to submit the most controversial chapters of the new book. Besides this, I was informed that even if *No Free Lunch* was accepted this side of the Atlantic, it might not be accepted with the Cambridge Syndicate in England, whose biologists were now disposed against my work. This news was actually quite surprising because the Cambridge Syndicate typically rubber stamps any recommendations for publication from the United States. That an exception might be made in my case indicated that the review process, instead of working dispassionately and fairly, would now be singling me out for special treatment. I therefore took my business elsewhere and published the book with Rowman and Littlefield.

My own experience with the peer-review process confirms an observation by Paul Gross: "Being right isn't enough. What you say, however right, must be said in a currently acceptable language, must not violate too brutally current taste, and must somehow signal your membership in a respectable professional club" (<www.mbl.edu/publications/Gross/Heilbrunn>). I was an unknown entity when I published *The Design Inference*, and the book didn't address the implications of the design inference for biology. Once those implications became clear, however, getting my work published in the peer-reviewed literature became more challenging—though certainly not impossible. I have, for instance, another book coming out with Cambridge University Press titled *Debating Design*, which I co-edited with Michael Ruse. Six out of seven referees approved it enthusiastically, and the lone dissenter grudgingly admitted that it would sell very well.

For research within an accepted framework, peer review is useful for quality control. But for radical new ideas and thinking outside the box, peer review is more often a hindrance than a help. That should come as no surprise given the nature of peer review. Peer review is primarily in the business of seeing that the standards, norms and practices of an established guild are respected. Only after they have been respected does the question of originality and innovation receive consideration. Peer review is essentially conservative. Peer review is therefore the last place we should expect to see a scientific revolution vindicated.

The history of peer review bears this out. As Frank Tipler has pointed out to me, the very idea of peer review as the touchstone for scientific truth and merit is a post-Second World War invention. In physics, for instance, peer-reviewed journals were not the norm until after 1950. In Germany, during the "Beautiful Years" (the period when quantum mechanics was being invented in the 1920s) one of the leading German physics journals, *Zeitschrift für Physik,* was not peer-reviewed: any member of the German Physical Society could publish there by simply submitting the paper.

Hence, if you had a really wild idea, all you had to do to get it published was ask a member of the German Physical Society to submit it for you. If you were a member, you could of course submit it for yourself. Werner Heisenberg published his paper on the uncertainty principle in this journal, and Alexandre Friedmann published his paper on the Friedmann universe (now the standard cosmological model) there as well. No peer review, just lots of brilliant physics.

All these observations about the history and nature of peer review are no doubt very interesting, but critics of intelligent design are unlikely to be impressed. It's all very well to say that design theorists publish work in the peer-reviewed literature that supports intelligent design. But intelligent design's main focus is biology. Are design theorists publishing and having work that supports intelligent design cited in the peer-reviewed *biological* literature? In other words, are we actually making inroads into mainstream biology? Design critics like Eugenie Scott, Paul Gross and Barbara Forrest will often state publicly that design theorists have published exactly *zero* articles in the peer-reviewed biological literature that support intelligent design.

Where do they get the number *zero?* I happen to have in front of me articles from the *Proceedings of the National Academy of Sciences, Journal of Molecular Biology, Journal of Theoretical Biology, Origins of Life and Evolution of the Biosphere* and *Annual Review of Genetics.* (The latter explicitly and favorably cites my book *No Free Lunch.*) They are all written by design theorists and are listed in the ISCID bibliography (<www.iscid.org/bibliography/bibliography.php>). And they all, in my view and that of the authors, support intelligent design. But that's just the problem. How can anything support intelligent design?

Critics like Eugenie Scott, Paul Gross and Barbara Forrest don't just deny that design theorists have published any works in the peer-reviewed literature that support intelligent design. They also deny that there could be any evidence at all that supports intelligent design. Yes, the articles I'm looking at are by design theorists. And yes, they are in the peer-reviewed biological literature. Yet according to critics, they can't support intelligent design. But is it that no evidence supports intelligent design? Or is it that plenty of evidence supports it provided that evidence is not ruled inadmissible on a priori grounds? Apriorism has an unhappy place in the history of science. For instance, science in Kepler's day knew that the orbits of the planets had to be circular. (Kepler's contemporary Galileo was adamant on this point.) Thus Kepler's evidence for elliptical orbits was ruled inadmissible because science "knew in advance" that the orbits had to be circular. In the end, however, Kepler was vindicated and the apriorist science of his day had to backpedal. That's always the danger with apriorism in science.

Critics of intelligent design who want to maintain that the number of articles in the peer-reviewed biological literature that support intelligent design is *zero* are playing a losing hand. That fiction is becoming increasingly difficult to maintain. Even so, I expect it to be maintained for a time. The problem is that to get work that supports intelligent design published in the peer-reviewed biological literature, biologists who are design theorists have to play their cards very close to the vest. As Michael Behe pointed out in an interview with the *Harvard Political Review* (<www.hpronline.org/news/251835.html>), for a biologist to question Darwinism endangers one's career: "There's good reason to be afraid. Even if you're not fired from your job, you will easily be passed over for promotions. I would strongly advise graduate students who are skeptical of Darwinian theory not to make their views known."

In the current intellectual climate it is impossible to get a paper published in the peer-reviewed biological literature if that paper explicitly affirms intelligent design or explicitly denies Darwinian and other forms of naturalistic evolution. Doubting Darwinian orthodoxy is comparable to opposing the party line of a Stalinist regime. What would you do if you were in Stalin's Russia and wanted to argue that Trofim Lysenko was wrong? You might point to paradoxes and tensions in Lysenko's theory of genetics, but you could not say that Lysenko was fundamentally wrong or offer an alternative that clearly contradicted Lysenko. That's the situation we're in. To get published in the peer-reviewed literature, design theorists have to tread cautiously and can't be too up front about where their work is leading. Indeed, that's why I was able to get Cambridge University Press to publish *The Design Inference* but not *No Free Lunch,* which was much more explicit in its biological implications.

By the way, you may be wondering why I don't here simply provide a list of peer-reviewed articles by design theorists from the biological literature that support intelligent design. The reason is that I want to spare these authors the harassment they would receive if I listed their work in this book. Overzealous critics of intelligent design regard it as their moral duty to keep biology free from intelligent design, even if that means taking extreme measures. I've known such critics to contact design theorists' employers and notify them of the "heretics" in their midst. Once "outed," the design theorists themselves get harassed and harangued with e-mails. Next, the press does a story mentioning their unsavory intelligent design associations. (The day one such story appeared, a close friend and colleague of mine mentioned in the story was dismissed from his research position at a prestigious molecular biology laboratory. He had worked in that lab for ten years.) Hereafter, the first thing that an Internet search of their names reveals is their connection with intelligent design. Welcome to the inquisition.

I close with one final point about peer review. Although intelligent design research is being published and cited in the peer-reviewed scientific literature (including the biological literature), even if it were not, that would not invalidate intelligent design. Lack of peer review has never barred the emergence of good science. Nor, for that matter, have peer-reviewed journals been the sole place where groundbreaking scientific work was done. As I noted, the peer-review process is inherently conservative, working nicely for filtering good incremental science from less rigorous work within an established paradigm, but it is lousy at opening its arms to paradigm revolutions. Thomas Kuhn, along with other eminent historians of science, has settled this point definitively: the old guard never opens its arms to a scientific revolution; they have too much invested in the old paradigm. The most important revolutions in science bypassed the peer-review process entirely and appeared in books. Copernicus's *De Revolutionibus,* Galileo's *On Two World Systems* and Newton's *Principia* are cases in point. None of these works were peer-reviewed. Nor was that book by a retiring English biologist from the nineteenth century—an unconventional work titled *On the Origin of Species.*

42

THE "WEDGE"

Isn't intelligent design really a political agenda masquerading as a scientific research program?

TWO ANIMATING PRINCIPLES DRIVE INTELLIGENT DESIGN. The more popular principle takes intelligent design as a tool for liberation from ideologies that suffocate the human spirit, such as reductionism and materialism. The other animating principle, less popular but intellectually more compelling, takes intelligent design as the key to opening up fresh insights into nature. The first of these animating principles is purely instrumental: it treats intelligent design as a tool for attaining some other end (like defeating materialism). Presumably, if other tools could more effectively accomplish that end, intelligent design would be abandoned. The second of these animating principles, by contrast, is intrinsic: it treats intelligent design as an essential good, an end in itself worthy to be pursued because of the insights it provides into nature.

These animating principles can work side by side, and there is no inherent conflict between them. Nonetheless, there is a clear order of priority. Unless intelligent design is an intrinsic good—unless it can be developed as a scientific research program and provide sound insights into the natural world—its use as an instrumental good for defeating ideologies that suffocate the human spirit becomes insupportable. Intelligent design must not become a "noble lie" for vanquishing views we find unacceptable. (History is full of noble lies that ended in disgrace.) Rather, intelligent design needs to convince us of its truth on its scientific merits. Then, because it is true and known to be true, it can serve as an instrument for liberation from suffocating ideologies—ideologies that suffocate not because they tell us the grim truth about ourselves, but because they are at once grim and false. Freud's psychic determinism is a case in point.

Intelligent design's dual role as a constructive scientific project and as a

means for cultural renaissance should raise some concerns over characterizing our movement as a "wedge." Intelligent design's instrumental good
of renewing culture hinges on its intrinsic good of furthering science. Unfortunately, the metaphor of the wedge clouds this order of precedence.
The wedge metaphor, as Phillip Johnson initially used it, focused on the
discrepancy between science as an empirical enterprise that goes where
the evidence leads (which is a legitimate conception of science) and science as applied materialist philosophy that maintains its materialism regardless of evidence (which is a bogus, though widely held, misconception of science). According to Johnson, the discrepancy between these two
conceptions of science provides a point of weakness into which the thin
end of a wedge can be inserted. Pounding the wedge at that point of weakness is supposed to invigorate science, renew culture and liberate society
from the miasma of materialism and naturalism. That's the promise.

Worthy goals though these are, their accomplishment is not appropriately ascribed to a wedge. Wedges break things rather than build them up.
Wedges are provisional and instrumental, driving toward some end but
not ends in themselves. Note the full title of Johnson's book: *The Wedge of
Truth: Splitting the Foundations of Naturalism.* I submit that the foundations
of naturalism are already split (thanks largely to Johnson's efforts). Even
now the right questions are on the table and being vigorously discussed.
What's more, the intelligent design movement is setting the terms (and
even the vocabulary) for the debate over biological origins. Karl Giberson
and Donald Yerxa (neither design advocates) make this point in *Species of
Origins,* which details the debate in the United States over evolution, creation and intelligent design:

> Since its inception in the early 1990s, the intelligent design move-
> ment has attracted so much attention that it has succeeded in domi-
> nating the origins debate. By this we do not mean that it is
> triumphant. Far from it. While design has made some modest in-
> roads in the academy, it is frequently seen . . . as a more attractively
> packaged variety of creationism. But design has succeeded in setting
> the agenda for much of the debate.

The wedge metaphor has outlived its usefulness. Indeed, with design
critics like Barbara Forrest and Paul Gross writing books like *Evolution and
the Wedge of Intelligent Design: The Trojan Horse Strategy,* the wedge metaphor has even become a liability. To be sure, critics of intelligent design
will attempt to keep using the wedge metaphor as a term of abuse. But the

wedge needs to be seen as a propaedeutic—as an anticipation of and prep-
aration for a positive, design-theoretic research program that invigorates
science and renews culture. The wedge, to vigorously mix metaphors, has
already swept the field, cleaned house, shone the spotlight and exposed
scientific materialism's dirty laundry. Now that that has been accomplished, where do we go from here?

Several observers watching intelligent design's progress as an intellectual movement have expressed concern about intelligent design being hi-
jacked as part of a larger cultural and political movement. In particular, intelligent design is thought to have been prematurely drawn into
discussions of public science education. First, such thinking goes, intelligent design needs to gain broad recognition from the scientific community
that it contributes significantly to our understanding of the natural world.
Bruce Gordon's widely quoted remarks in the January 2001 issue of *Research News and Opportunities in Science and Theology,* for instance, make
this point.

There is an important kernel of truth in these concerns, but only a kernel. Intelligent design does need to succeed as a scientific enterprise to succeed as a cultural and political enterprise. In other words, the instrumental
good of intelligent design cannot race ahead untethered from the progress
of its intrinsic good. But this is quite different from requiring that intelligent design gain broad recognition in the scientific world before it may be
regarded as a bona fide intellectual project and legitimately influence public opinion and policy.

Intelligent design's legitimacy as an intellectual project hinges on two
facts that are independent of its recognition or acceptance within the scientific world. First, evolutionary biology has been so hugely unsuccessful
as a scientific theory in accounting for the origin of life and the emergence
of biological complexity that it does not deserve a monopoly regardless
what state of formation intelligent design has reached. Second, intelligent
design is, logically speaking, the only alternative to a mechanistic evolutionary biology (see chapter thirty-six). Evolutionary biology, as currently
formulated, embraces material mechanisms and eschews teleology. Yet
these are the only two available options: either material mechanisms can
do all the work in biological origins or some telic process is additionally
required. The issue before the public square, therefore, is not in the first instance how far intelligent design has developed as a scientific project but
whether it will be granted freedom and equity. In particular, are all sectors
of the public free to examine and discuss the full range of scientific options

concerning biological origins? Design theorists say yes. Darwinism's defenders prefer that certain sectors of the public (like public education) be cordoned off and censored. What's more, to justify their censorship, they employ spurious, question-begging definitions of science and rationality.

Any setting of rules about what intelligent design must accomplish in the scientific sphere before it may legitimately influence the political sphere is arbitrary and betrays either a naiveté about the actual workings of science or a shrewdness about how scientific enterprises get funded and promoted in the public sphere (and thus, conversely, how to undermine the success of a promising but undesired scientific enterprise). What a clean, happy world it would be if every power struggle were played out fairly, gently and with due respect for the opponent. All mature men and women know that this is rarely the case. In fact, scientific materialists set unreasonable rules not because they are incapable of reasonable thought but because such rule setting quite intentionally undermines intelligent design research from going forward. For a scientific research program to prosper, it must employ talented workers and ensure that their efforts to further the program get rewarded. This requires societal and political structures to be in place that can attract talented workers and offer them incentives for a fruitful career. Science, culture, society and politics all work together in this regard. Insofar as Darwinists hold the keys to these bank vaults and cultural resources, they are none too eager to share them.

Although intelligent design as a scientific program stands logically prior to intelligent design as a cultural movement, this logical priority does not imply temporal priority. To think that the scientific program must first succeed according to some arbitrary criterion of success (the Darwinists' criterion?) before the cultural movement can legitimately be undertaken is not only naïve but to give up on both. The scientific research and cultural renewal aspects of intelligent design need to work together, protecting and reinforcing each other. Science grows within a cultural matrix but at the same time shapes that matrix. Their relation is not linear but dialectical. Intelligent design advocates have openly and methodically pursued these dialectically related ends and have encouraged its opponents to engage intelligent design in the same spirit—not with rule setting, name calling, deck stacking or character assassination, but openly and methodically, trusting that truth, as John Milton argued, will rise to the top through a free and open exchange in the marketplace of ideas.

43

RESEARCH THEMES

*What's a scientist interested in
intelligent design supposed to do by
way of scientific research?*

PLENTY OF SCIENTISTS ARE INTRIGUED WITH intelligent design but for now don't see how they can contribute usefully to it. I recently had an exchange with one such scientist (a geneticist who is also a Christian). I asked him, "What sort of real work needs to go forward before you felt comfortable with intelligent design?" His response was revealing: "If I knew how to scientifically approach the question you pose, I would quit all that I am doing right now, and devote the rest of my career in pursuit of its answer. The fact that I have no idea how to begin gathering scientific data that would engage the scientific community is the very reason that I don't share your optimism that this approach will work."

Or consider Francis Collins, head of the Human Genome Project. As a Christian believer, he is committed to design in some broad sense. Yet at a meeting of the American Scientific Affiliation (at Pepperdine University, August 2-5, 2002), he expressed doubts about intelligent design as a scientific project. The problem, according to him, is intelligent design's "lack of a plan for experimental verification."

I remain supremely optimistic that intelligent design has the research potential to satisfy such scientists. That potential, however, needs to be actualized. How to actualize it? The most important thing right now is a steady stream of good ideas together with the resources to implement them. In particular, we need to reflect deeply about biological systems. That reflection needs to generate profound insight. And that insight needs to get us asking interesting new questions that can be framed as research problems. With these research problems in hand, we then need to go to nature and see how they resolve.

I'm mainly a theoretician, so I'm not in a position to lay out a detailed set of research problems for intelligent design. Nonetheless, as an interdisciplinary scholar who rubs shoulders with scientists from many disciplines, I am in a position to lay out some *research themes* that may prove helpful to scientists who are trying to find a way to contribute productively to intelligent design research. What follows, then, is a list of research themes. (Let me stress that I make no pretense at completeness.)

1. *Design detection.* Techniques, methods and criteria of design detection are widely employed in various special sciences (such as forensics, archeology, cryptography and the search for extraterrestrial intelligence, or SETI). There's currently much discussion from all sides about the validity of detecting actual design in biology using Michael Behe's criterion of irreducible complexity or my criterion of specified complexity. Design theorists need to be at the center of this discussion.

2. *Biological information.* The word *information*, according to its Latin etymology, means "to give shape or form" to something. It's no exaggeration to say that the origin of life and its subsequent complexification constitutes an "information revolution" in the history of matter. Indeed, matter needs to be formed in very special ways in order to constitute life. What is the nature of biological information? How do function and fitness relate to it? What are the obstacles that face material mechanisms in attempting to generate biological information? Most importantly, what are the theoretical and empirical grounds for thinking that intelligence is indispensable to the origin of biological information? I've begun to address these problems in my book *No Free Lunch,* but much more work is needed here.

3. *Minimal complexity.* Living things are complex systems that consist of complex subsystems that in turn consist of complex subsubsystems and so on until a level of organization is reached that is chemically simple (for instance, individual amino acids or nucleotide bases). How does pruning away the complexity of such systems affect their ability to perform some function or set of functions (notably, keeping the organism alive and able to reproduce)? How much complexity can be pruned away without losing function? Once a complexity barrier is reached below which function can no longer be preserved, could coevolution overcome that barrier by switching function? Are there systems that not only are minimally complex with respect to some function, but for which any reduction of complexity eliminates all possibility of biological function? Would such systems provide decisive confirmation of intelligent design?

4. *Evolvability.* Evolutionary biologists are in the business of drawing evolutionary connections between biological systems. This requires identifying biological systems, relating them according to some similarity metric and then telling evolutionary stories that connect the dots. Yet for large-scale evolutionary changes, these stories tend to be imaginative reconstructions for which evidence is thin to nonexistent. This is certainly true of attempts to bridge major divisions in the fossil record. It is also true of molecular phylogenies. Evolutionary biology's preferred research strategy consists in taking distinct biological systems and trying to merge them. Intelligent design, by contrast, focuses on a different strategy, namely, taking individual biological systems and perturbing them to see how much the systems can evolve (with and without intelligence). Limitations on evolvability by material mechanisms constitute evidence for design.

5. *The principle of methodological engineering.* The reason evolutionary biology has lost all sense of proportion about how much evolution is possible as a result of blind material mechanisms (like random variation and natural selection) is that it floats free of the science of engineering. At every crucial juncture where some major evolutionary transition needs to be accounted for, evolutionary biology invokes a designer-substitute (such as natural selection, lateral gene transfer or symbiogenesis) to do the necessary design work. Yet unlike the science of engineering, evolutionary biology does not actually perform the necessary design work or specify a detailed procedure by which it might be accomplished. Intelligent design, by contrast, takes what I call "methodological engineering" as a fundamental regulative principle for understanding biological systems. According to this principle, biological systems are to be understood as engineering systems. In consequence, their origin, construction, operation, break down, wearing out, repair and, above all, history of modifications (both designed and accidental) are all to be understood in engineering terms. In the next ten years I foresee academic programs in biotic engineering supplanting academic programs in evolutionary biology.

6. *Technological evolution (TRIZ).* The only well-documented example we have of the evolution of complex multipart integrated functional systems (as we see in biology) is the technological evolution of human inventions. In the second half of the twentieth century, Russian scientists and engineers studied hundreds of thousands of patents to determine how technologies evolve. They codified their findings in a theory to which they gave the acronym TRIZ, which corresponds to a Russian phrase that in English means "theory of inventive problem solving" (and is sometimes given the acronym TIPS). The picture of technological evolution that

emerges out of TRIZ maps amazingly well onto the history of life as we see it in the fossil record and includes the following:

- New technologies (cf. major groups like phyla and classes) emerge suddenly as solutions to inventive problems. Such solutions require major conceptual leaps (i.e., design).

- Existing technologies (cf. species and genera) can, by contrast, be modified by trial-and-error tinkering (cf. Darwinian evolution), which amounts to solving routine rather than inventive problems. (The distinction between routine and inventive problems is central to TRIZ. In biology, irreducible complexity suggests one way of making the analytic cut between these types of problems. Are there other ways?)

- Technologies approach ideality (cf. local optimization by means of natural selection) and thereafter tend not to change (cf. stasis).

- New technologies, by supplanting old technologies, can upset the ideality and stasis of the old technologies, thus forcing them to evolve in new directions (requiring the solution of new inventive problems, as in an arms race) or by driving them to extinction.

Mapping TRIZ onto biological evolution provides a potentially fruitful avenue of design-theoretic research that is entirely consonant with the principle of methodological engineering.

I need here to add a footnote about TRIZ. Most design critics, by conflating intelligent design with creationism, see intelligent design as committed to a designer who always designs from scratch and has to get everything right the first time. TRIZ, by contrast, bespeaks an evolutionary process that as much as possible takes advantage of existing designs but then at key moments requires a conceptual breakthrough to move the process of technological evolution along. On this view, the process of technological evolution is itself designed. What's more, within that process, designing intelligences interact with natural forces. Does this mean that the designer (or designers) is making things up as it goes along? Not necessarily. The conceptual breakthroughs needed to drive technological evolution can be programmed from the start. And what about suboptimal and dysteleological design? These can be explained in part as the result of natural forces subverting an original design plan. Teasing apart the effects of intelligence from natural forces thus becomes a key research question for a TRIZ approach to intelligent design.

7. Autonomy versus guidance. Many scientists worry that intelligent design attempts to usurp nature's autonomy. But that's not the case. Intelligent design is merely trying to restore a proper balance between nature's autonomy and teleologic guidance. Prior to the rise of modern science all the emphasis was on teleologic guidance (in the form of divine design). Now the pendulum has swung to the opposite extreme, and all the emphasis is on nature's autonomy (an absolute autonomy that excludes design). Might there not be a midpoint that properly respects both and at which design becomes empirically evident? The search for that midpoint needs always to be in the back of our minds as we engage in design-theoretic research. It's not all design or all nature but a synergy of the two. Unpacking that synergy is the intelligent design research program in a nutshell.

8. *Evolutionary computation*. It is becoming increasingly evident that organisms employ evolutionary computation to solve many of the tasks of living. But does this show that organisms originated through some form of evolutionary computation (as through a Darwinian evolutionary process)? It seems that the immune system, for instance, is a general-purpose genetic algorithm that targets an interloper, sets up a gradient that tracks the interloper and then runs a genetic algorithm specifically adapted to that gradient whose output is a molecular assemblage that vanquishes the interloper. All of this sounds very high-tech and programmed. Are GPGAs (general-purpose genetic algorithms) like this actually designed, or themselves the result of evolutionary computation?

Evolutionary computation occurs in the behavioral repertoire of organisms but is also used to account for the origination of certain features of organisms. It would be helpful to explore the relationship between these two types of evolutionary computation as well as any design intrinsic to them. My work in chapter four of *No Free Lunch* lays out some of the theoretical groundwork for this. We need, in addition to theoretical work in this area, a large contingent of design-theorist computer programmers to write and run computational simulations that investigate the scope and limits of evolutionary computation. One such simulation is the MESA program (Monotonic Evolutionary Simulation Algorithm) due to Micah Sparacio, John Bracht and myself. It is available on the ISCID website (<www.iscid.org/mesa>).

9. *Understanding discontinuity*. Evolution is committed to continuity in a broad sense. Its main business is to connect dots. But for dots to be plausibly connected, they need to be reasonably close together. That's why the absence of transitional forms, gaps and missing links or intermediates constitute a problem for evolution. To be sure, evolutionists do not regard the absence of intermediates as a problem in the bad sense. They regard such discontinuities not as challenges to their theory but as discontinuities that are only apparent and that will disappear once the missing intermediates are found. Consequently, whenever an intermediate is found, it is regarded as a triumph for evolutionary theory. (Witness the recent excitement over the Toumaï fossil find in Chad.)

Evolutionary biology attempts to explain the absence of intermediates from an evolutionary path on the assumption that the intermediates did once exist. But let's turn the question around. Suppose that discontinuity is a fact not just about the history of life as we know it but about the history of life itself: in other words, suppose the intermediates never existed. In that case, how did biological forms in all their vast complexity and diversity come about? In asking this question, let's hold off asking for the underlying cause or causes of biological complexity and diversity. Rather, let's merely ask what a video camera would see if it were scouring the past and recording key events in life's history. There are exactly four possibilities:

- *Nonbiogenic emergence:* Organisms emerge without the direct causal agency of other organisms. In place of life begetting life, here we have nonlife begetting life.

- *Generative transmutation:* Organisms, in reproducing, produce offspring that are vastly different from themselves.

- *Biogenic reinvention:* Organisms reinvent themselves in midstream. At one moment they have certain morphological and genetic features; at the next they have a vastly different set of such features.

- *Symbiogenic reorganization:* Organisms emerge when different organisms from different species get together and reorganize themselves into a new organism.

None of these possibilities is out to lunch. Nonbiogenic emergence had to happen at least once, namely, at the origin of life. Symbiogenic reorganization has been Lynn Margulis's main focus of research, and there is increasing evidence for it. Biogenic reinvention (organisms changing in midstream) is also not that crazy when one considers the life cycles of certain organisms that from one stage to the next are completely unrecognizable (e.g., the metamorphosis of the butterfly or, even more extremely, the various stages of the liver fluke). Finally, generative transmutation suggests a programmed view of evolution where, like a computer program that kicks in at a certain time (recall the Michelangelo computer virus that kicked in March 6, 1993), organisms change in one generation. French paleontologist Anne Dambricourt has argued for this view in respect to the emergence of *Homo sapiens*. With regard to these four possibilities, the crucial question is how to make sense of them in light of intelligent design. Clearly, none of them makes sense without massive coordination of chemical and biological processes, a coordination that bespeaks intelligent direction.

10. Steganography. Finally, we come to the research theme that I find most intriguing. *Steganography,* according to the dictionary, is an archaic expression that was subsequently replaced by the term *cryptography*. Steganography literally means "covered writing." With the rise of digital computing, however, the term has taken on new life. Steganography belongs to the field of digital data embedding technologies (DDET), which also include information hiding, steganalysis, watermarking, embedded data extraction and digital data forensics. Steganography seeks algorithms that are efficient (i.e., have a high data rate) and robust (i.e., are insensitive to common distortions) and that can embed a high volume of hidden message bits within a cover message (typically imagery, video or audio) without their presence being detected. Conversely, steganalysis seeks statistical tests that will detect the presence of steganography in a cover message.

Consider now the following possibility: What if organisms instantiate designs that have no functional significance but that nonetheless give biological investigators insight into functional aspects of organisms? Such second-order designs would serve essentially as an "operating manual"— of no use to the organism as such but of use to scientists investigating the organism. Granted, this is a speculative possibility, but there are some preliminary results from the bioinformatics literature that bear it out in relation to the protein-folding problem. (Such second-order designs appear to be embedded not in a single genome but in a database of homologous genomes from related organisms.)

While it makes perfect sense for a designer to throw in an "operating manual" (much as automobile manufacturers include operating manuals with the cars they make), this possibility makes no sense for blind material mechanisms, which cannot anticipate scientific investigators. Research in this area would consist in constructing statistical tests to detect such second-order designs (in other words, steganalysis). Should such second-order designs be discovered, the next step would be to identify algorithms for embedding these second-order designs in the organisms. My suspicion is that biological systems do steganography much better than we do and that steganographers will learn a thing or two from biology—though not because natural selection is so clever, but because the designer of these systems is so adept at steganography.

Such second-order steganography would, in my view, provide decisive confirmation for intelligent design. What's more, its investigation would be entirely reputable in a way that biblical numerology and Bible codes, say, are not. The problem with biblical numerology and Bible codes is that the Bible has a plain sense (i.e., the meaning conveyed by the actual words of Scripture). Biblical numerology and Bible codes, however, postulate a hidden sense of the Scripture. Yet to postulate such a hidden sense is to claim that the Bible contains hidden information that could just as well have been contained in the plain sense.

Second-order steganography is not like this. Steganographic information that's useful only to a scientific investigator but not to an organism would be different in kind from the functional information needed by an organism to build complex structures independent of such investigators. The information hidden within biblical numerology and Bible codes could have been included in the Bible's plain sense. Second-order steganographic information, by contrast, could not be included in the organism's functional information because these forms of information are intended for two completely different audiences: scientific investigators and organisms (i.e., scientific investigators with their need to understand nature in the one case, organisms with their need to survive and reproduce in the other).

Even if second-order steganography doesn't pan out, first-order steganography (i.e., the embedding of functional information useful to the organism rather than to a scientific investigator) could also provide strong evidence for intelligent design. For years now evolutionary biologists have told us that the bulk of genomes is junk and that this is due to the sloppiness of the evolutionary process. That is now changing. For instance, researchers at the University of California at San Diego are finding that long stretches of seemingly barren DNA sequences may form a new class of noncoding RNA genes scattered, perhaps densely, throughout animal genomes. Design theorists should be at the forefront in unpacking the information contained within biological systems. If these systems are designed, we can expect the information to be densely packed and multilayered (save where natural forces or intentional disruption have attenuated the information). Dense, multilayered embedding of information is a prediction of intelligent design.

44

MAKING INTELLIGENT DESIGN
A DISCIPLINED SCIENCE

*Granting that intelligent design is a
scientific research program, or that it at
least has the potential to become one, how
can it avoid being swept away as part of a
larger cultural and political agenda?*

INTELLIGENT DESIGN HAS MADE TREMENDOUS inroads into the culture at
large. Front-page stories featuring the work of design theorists have appeared in the *New York Times, Los Angeles Times, Wall Street Journal, San Francisco Chronicle* and so on. Television, radio and weeklies like *Time* magazine
have focused the spotlight on intelligent design as well. This publicity is at
once useful and seductive. It is useful because it helps get the word out and
attract talent to the movement. It is seductive because it can deceive us into
thinking that we have accomplished more than we actually have.

Although proponents of intelligent design have done amazingly well in
creating a cultural movement, we must not overstate intelligent design's
successes on the scientific front. It is fine to receive respectful notice from
the *New York Times.* But, as David Berlinski has pointed out to me, René
Thom's catastrophe theory also received front-page coverage in the *Times;*
and thereafter, despite its real content as both a scientific theory and a
philosophical attitude, it died quietly some time in the 1980s. An intellectual movement cannot sustain itself on media attention. The scientific and
conceptual work on intelligent design occurs out of the limelight, requires
intense concentration over extended periods and is fully appreciated only
by relatively few specialists. The cultural renewal work on intelligent design, by contrast, occurs in the limelight, offers quick closure and gratifica-
tion and makes its appeal to the population at large.

We need to be very clear when we are doing the nuts-and-bolts scientific and conceptual work on intelligent design and when we are engaged in cultural and political activity. What's more, these aspects of intelligent design need to keep pace. Because of intelligent design's outstanding success at gaining a cultural hearing, the cultural and political component of intelligent design is now running ahead of the scientific and intellectual component. I want therefore to lay out a series of recommendations for rectifying this imbalance.

1. Catalog of fundamental facts. One of the marks of a disciplined science is that it possesses an easily accessible catalog of fundamental facts. Think of the magnificent star cluster catalogs in astrophysics. Intelligent design needs something like this. It would be enormously helpful if we had, and could make publicly available, a catalog of irreducibly complex biological objects or processes. The catalog should contain as complete a list as possible, organized more or less as a table, with very complete descriptions. Under the bacterial flagellum, for instance, the catalog would list these: found in the following; involving these biochemical parts; requiring this level of energy; utilizing these substrates; etc. The catalog should move from simple to profound examples of irreducible complexity (such as the mammalian visual system).

The criteria governing entries should be very strict and should be stated explicitly: *such and such* is irreducibly complex if and only if *fill-in-the-blank*. The catalog should be widely distributed to the biological community. It should have no mention of intelligent design, nothing about naturalism—just the fundamental facts as they are now known. Such a catalog would do more than any number of forums or debates to persuade biologists that Darwinism is in trouble and that intelligent design is a live possibility. Right now most of them don't even see that there is a problem. Irreducible complexity is for them not a problem urgently in need of resolution but a detail to be shelved indefinitely. Such a catalog would put an end to the current complacency.

2. Catalog correcting misinformation. There is a tremendous amount of misinformation in the biological literature whenever it impinges on design. Jonathan Wells's *Icons of Evolution* is an attempt to redress that problem by examining a few faulty evidences used to prop up evolutionary theory at the expense of design. But the problem is pervasive. Sometimes it's merely giving an evolutionary spin to a biological experiment or fact when the actual evidence warrants nothing of the sort. Sometimes it's the double standard by which natural selection gets applied: if a biological system looks well designed, that's because natural selection is an efficient designer-substitute that prunes away deadwood; on the other hand, if it looks cobbled together, that's because natural selection is a sloppy opportunist.

The suboptimality objection has traditionally been Darwinism's ace for keeping intelligent design at bay. But as with so many protective measures, it ends up undermining the very thing it was designed to protect, namely, a true understanding of biological systems. To refute design, critics resort to belittling systems they claim are not designed (the logic from incompetent design to no design presumably being *de rigueur*). In repudiating design, biologists therefore consistently underestimate biological systems.

The catalog I am proposing would document as much. Like the previous catalog, this catalog is not optional. Biology is firmly in the grip of an anti-design bigotry that needs to be unmasked and defeated. As David Berlinski has put it to me, "A shift in prevailing scientific orthodoxies will come only when the objections to Darwinism . . . accumulate so forcefully that they can no longer be ignored."

3. Network of researchers and resources. Intelligent design as a scientific and intellectual project has many sympathizers but few workers. The scholarly side of the movement at this time consists of a handful of academics and independent researchers. These numbers need to swell, and we need to get properly networked. We need to know who's out there working on what. To this end the Internet will prove invaluable. Intelligent design is at this time still an academic pariah. Consequently, it is difficult to concentrate our forces in any one institution. And yet, when I speak about intelligent design on university campuses, I almost invariably encounter at least one scientist on faculty eager to do research on intelligent design. The Internet, particularly as live chats and videoconferences become more readily available, will bring together scholars who now work in isolation. This will help overcome the institutional barriers they now face. Full and effective use of the Internet is simply a must.

The natural place to house such a network is within a professional society. Fortunately, such a society is now in place—the International Society for Complexity, Information and Design (<www.iscid.org>). Housing the network at this site is an option, though there are other options. The important thing for now is that we get networked, not who does the networking. Associated with this network should be *research coordinators* expert in a given field of science to help researchers in that field coordinate their efforts. The network needs to be endowed with resources. ISCID currently offers an annotated bibliography of design-relevant literature. Access to various online subscription services (journals, specialized search engines, electronic books, etc.) should also be part of the resource package. This will cost money, but it will be well worth the investment. Concentration of forces is a key principle of military tactics. Without it, troops, though willing and eager, wallow in indecision and cannot act effectively. The network of researchers and resources that I am recommending is the first step in concentrating our forces. The next step is setting the intellectual agenda for academic departments and even whole academic institutions. But that is still downstream and will depend on the next recommendation.

4. *Building a design curriculum.* Ivan Pavlov and John Watson were both active in the early part of the twentieth century developing a behaviorist psychology. Behaviorism itself, however, didn't take off as an intellectual movement until a generation later, when psychologists built a curriculum around it. For scientific ideas to prosper (regardless of whether they are correct or ultimately mistaken, as behaviorism proved to be), they must be part of a curriculum that gets taught within the educational mainstream. This is the only way to win the next generation of scholars to intelligent design. Without a presence in the science curriculum, intelligent design will limp along, merely winning stragglers here and there.

A problem we now face with intelligent design is that even if the educational mainstream opened its arms to us (don't hold your breath), we have no sustained course of study to give them. A curriculum provides that and much more. It takes the crazy-quilt of science and systematizes it into an intellectually coherent position. Students are thus introduced to a research tradition and not merely to a disconnected set of claims and arguments, or worse yet to some effective but easily ignored criticisms. Darwinists, by contrast, have a curriculum—indeed, one that is steadily gobbling up discipline after discipline (evolutionary psychology being one of the more visible recent additions). Daniel Dennett was right when he called Darwinism a universal acid. Darwinism's hold on the academy is pervasive and monopolistic. By building a design curriculum, we attempt to restore a free market.

Are we at this time in the position to build a design curriculum? Certainly intelligent design as a scientific program needs to develop and mature. Nevertheless, I believe we are in a position to start building such a curriculum. At the very top of the list we need an introductory basal biology textbook—in other words, a standard 800- to 1,000-page introductory biology text framed around intelligent design rather than Darwinian evolution. Note that such a text would provide a fair and detailed treatment of Darwinian evolution. In fact, it would tell students more about Darwinian evolution than Darwinists typically want them to hear, notably about the theory's problems and weaknesses. (And we don't even need to cite ourselves here. Critics within evolutionary biology's own ranks, like the late Stephen Jay Gould and now Lynn Margulis with her theory of symbiogenesis, have saved us the trouble.)

Actually, we'll need two basal biology texts, one geared toward college students and another simplified version geared toward high school students. The closest thing we have right now is a supplemental biology text (*Of Pandas and People* by Dean Kenyon and Percival Davis). This is a terrific book. Nevertheless, as a supplemental text, its market and readership is necessarily limited. Once we have a basal biology text, we need to go through each discipline where Darwinian and naturalistic thinking has been used to illegitimately exclude intelligent design. Darwinism's universal acid has eaten into many disciplines, ranging from the sciences to the humanities. To counteract that acid, design theorists need to target each such discipline and systematically rethink it. This work of reconceptualization and restoration will be very labor intensive and will require the efforts of many scholars. The disciplines at the top of the list (after biology) that need to be reconceptualized are evolutionary psychology, bioethics, cognitive neuroscience, artificial intelligence, philosophy of mind (especially the problem of consciousness), the history and philosophy of science, foundations of physics, and cosmology.

Building a design curriculum is educational in the broadest sense. It includes not just textbooks but everything from research monographs for professors and graduate students to coloring books for preschoolers. It needs to take full advantage of the technologies and media at our disposal: CD-ROMs, videos, DVDs, computer animation, e-learning and more. The videos *Unlocking the Mystery of Life* and *Icons of Evolution* are exemplary in this regard (available at <www.arn.org>). So too is the cartoon book *What's Darwin Got to Do with It?* (by Robert Newman and John Wiester), which provides a perfect lead-in for students about to study high school biology.

Martin Luther once remarked that we can do without lots of things, but we can't do without schools, for they must rule the world. Not only must they rule the world, but they do indeed rule the world. Without a significant presence in the educational mainstream, intelligent design will continue to be marginalized and will never attain its full potential. A design-theoretic curriculum is therefore indispensable to the success of intelligent design as a scientific and intellectual movement.

5. Objective measures of progress. How do we gauge how well we are doing in developing intelligent design as a scientific research program? We need some objective measures of progress. Rather than lay out such measures in pedantic detail, let me indicate what they are under four rubrics, each followed by a series of questions:

- *Intellectual vitality:* Have we become boring? Have we run out of things to say? Is the fount of fresh ideas drying up? Are we constantly repeating ourselves? Are people who once were excited about what we're doing no longer excited? Or do we have the intellectual initiative? Are we setting the agenda for the problems being discussed? Are we ourselves energized by our research? Is there nothing else we'd rather be doing than working on intelligent design? Are our ideas strong enough to engage the best and the brightest on the other side?

- *Intellectual standards:* Are we holding ourselves to high intellectual standards? Are we in the least self-critical about our work? Are we sober or immodest about our work? Do we demand precision and rigor from each other? Do we examine each other's work with intense critical scrutiny and speak our minds freely in assessing it? Or do we try to keep all our interactions civil and diplomatic (perhaps so as not to give the appearance of dissension in our ranks)? Does the mood of our movement alternate between the smug and the indignant—smug when we hold the upper hand, indignant when we are criticized? Do we react to adverse criticism like first-time novelists who are dismayed to discover that their "masterpiece" has been trashed by the critics? Or do we take adverse criticism as an occasion for tightening and improving our work?

- *Exiting the ghetto:* Do we refuse to be marginalized within an intellectual ghetto or second-class subculture? Are scholars and scientists on the other side actually getting to know us? Once they get to know us, do they still demonize us? Or do they think that we have an interesting, albeit perverse, point of view? Is intelligent design's appeal international? Does it cross religious boundaries? Or is it increasingly confined to American evangelicalism? Who owns intelligent design? Are we trying to get our ideas into the scientific mainstream? Are we continuing to plug away at getting our work published in the mainstream peer-reviewed literature (despite the deck being stacked against us)? Or are we seeking safe havens where we can publish our work easily, yet mainly for the benefit of each other? At ISCID, for instance, we encourage contributors to the society's journal also to submit their articles to the main-stream literature and are delighted when they get accepted there.

- *Attracting talent.* Are we continually attracting new talent to intelligent design's scientific research program? Does that talent include intellects of the highest caliber? Is that talent distributed across the disciplines or confined only to certain disciplines? Are underrepresented disciplines getting filled? What about talent that's been with the movement in the past? Is it staying with the movement or becoming disillusioned and aligning itself elsewhere? Do the same names associated with intelligent design keep coming up in print or are we constantly adding new names? Are we fun to be around? Do we have a colorful assortment of characters? Other things being equal, would you rather party with a design theorist or a Darwinist?

These, then, are my recommendations for turning intelligent design into a disciplined science. Will intelligent design implement these recommendations and thereby succeed as a scientific program, intellectual project and cultural movement? From the vantage of the scientific establishment, intelligent design is in the position of a mouse trying to move an elephant by nibbling at its toes. From time to time the elephant may shift its feet, but nothing like real movement or a fundamental change is about to happen. Yet despite taking this view, the scientific establishment seems strangely uncomfortable. The mouse has yet to be squashed, and the elephant (as in the cartoons) has become frightened and seems ready to stampede in a panic.

The image that I think more accurately captures how intelligent design will play out is, ironically, that of an evolutionary competition where two organisms vie to dominate an ecological niche. (Think of the mammals displacing the dinosaurs.) At some point one of the organisms gains a crucial advantage. This enables it to outcompete the other. One thrives, the other dwindles. However wrong Darwin might have been about exalting selection and competition as the driving force behind biological evolution, these factors certainly play a crucial role in scientific progress. It's up to intelligent design proponents to demonstrate a few incontrovertible instances where design is uniquely fruitful for biology. Scientists without an inordinate attachment to Darwinian evolution (and there are many, though this fact is not widely advertised) will be only too happy to shift their allegiance if they think that intelligent design is where the interesting problems in biology lie.

I close this chapter as well as this book with a quotation by Emanuel Lasker, philosopher, mathematician, friend of Albert Einstein and world chess champion for twenty-seven years. Strictly speaking, his comments are about chess. But for Lasker, chess was life and life was chess. Victory in chess was for him a triumph of truth. I present Lasker's quotation because he puts his finger on the honesty, precision and critical sense that must guide our thinking if we are to meet the challenges of evolutionary biology and turn intelligent design into a disciplined science. Here is the quotation, which is taken from *Lasker's Manual of Chess:*

> Life is generated only by life. He who wants to educate himself in Chess must evade what is dead in Chess—artificial theories, supported by few instances and unheld by an excess of human wit; the habit of playing with inferior opponents; the custom of avoiding difficult tasks; the weakness of uncritically taking over variations or rules discovered by others; the vanity which is self-sufficient; the incapacity for admitting mistakes; in brief, everything that leads to a standstill or to anarchy.

SELECT BIBLIOGRAPHY

Arnhart, Larry, Michael J. Behe and William A. Dembski. "Conservatives, Darwin & Design: An Exchange." *First Things,* November 2000, pp. 23–31.

Axe, Douglas. "Extreme Functional Sensitivity to Conservative Amino Acid Changes on Enzyme Exteriors." *Journal of Molecular Biology* 301 (2000): 585–95.

Ayala, Francisco J. "Darwin's Revolution." In *Creative Evolution?!* Edited by J. H. Campbell and J. W. Schopf. Boston: Jones and Bartlett, 1994.

Babbage, Charles. *The Ninth Bridgewater Treatise.* London: Murray, 1836.

Beckwith, Frank. *Law, Darwinism and Public Education: The Establishment Clause and the Challenge of Intelligent Design.* Lanham, Md.: Rowman and Littlefield, 2003.

Behe, Michael J. *Darwin's Black Box: The Biochemical Challenge to Evolution.* New York: Free Press, 1996.

Berlinski, David. "The Deniable Darwin." *Commentary,* June 1996, pp. 19–29.

———. "Has Darwin Met His Match?" *Commentary,* December 2002, pp. 31–41.

Borel, Emile. *Probabilities and Life.* Translated by M. Baudin. New York: Dover, 1962.

Campbell, John Angus, and Stephen C. Meyer, eds. *Darwinism, Design, and Public Education.* Lansing: Michigan State University Press, 2003.

Cicero, Marcus Tullius. *De Natura Deorum.* Translated by H. Rackham. Cambridge, Mass.: Harvard University Press, 1933.

Crick, Francis. *What Mad Pursuit: A Personal View of Scientific Discovery.* New York: BasicBooks, 1988.

Crick, Francis, and Leslie E. Orgel. "Directed Panspermia." *Icarus* 19 (1973): 341–46.

Dam, Kenneth W., and Herbert S. Lin, eds. *Cryptography's Role in Securing the Information Society.* Washington, D.C.: National Academy Press, 1996.

Darwin, Charles. *On the Origin of Species.* Facsimile 1st ed. 1859. Reprint, Cambridge, Mass.: Harvard University Press, 1964.

Davies, Paul. *The Fifth Miracle: The Search for the Origin and Meaning of Life.* New York: Simon & Schuster, 1999.

Davis, Percival, and Dean Kenyon. *Of Pandas and People.* 2nd ed. Dallas: Haughton, 1993.

Dawkins, Richard. *The Blind Watchmaker: Why the Evidence of Evolution Reveals a Universe Without Design.* New York: Norton, 1987.

———. *Climbing Mount Improbable.* New York: Norton, 1996.

Dembski, William A. *The Design Inference: Eliminating Chance Through Small Probabilities.* Cambridge: Cambridge University Press, 1998.

———. *Intelligent Design: The Bridge Between Science and Theology.* Downers Grove, Ill.: InterVarsity Press, 1999.

———. *No Free Lunch: Why Specified Complexity Cannot Be Purchased Without In-*

telligence. Lanham, Md.: Rowman and Littlefield, 2002.

―――――. "Randomness by Design." *Nous* 25, no. 1 (1991): 75–106.

Dembski, William A., ed. *Uncommon Dissent: Intellectuals Who Find Darwinism Unconvincing.* Wilmington, Del.: ISI Books, 2004.

Dembski, William A., and Michael Ruse, eds. *Debating Design: From Darwin to DNA.* Cambridge: Cambridge University Press, 2004.

Dennett, Daniel. *Darwin's Dangerous Idea.* New York: Simon & Schuster, 1995.

Denton, Michael. *Evolution: A Theory in Crisis.* Bethesda, Md.: Adler & Adler, 1985.

―――――. *Nature's Destiny: How the Laws of Biology Reveal Purpose in the Universe.* New York: Free Press, 1998.

Denton, Michael, J. C. Marshall and M. Legge. "The Protein Folds as Platonic Forms: New Support for the Pre-Darwinian Conception of Evolution by Natural Law." *Journal of Theoretical Biology* 219 (2002): 325–42.

Devlin, Keith. *Logic and Information.* Cambridge: Cambridge University Press, 1991.

Dretske, Fred. *Knowledge and the Flow of Information.* Cambridge, Mass.: MIT Press, 1981.

Earman, John. *Bayes or Bust? A Critical Examination of Bayesian Confirmation Theory.* Cambridge, Mass.: MIT Press, 1992.

Eigen, Manfred. *Steps Towards Life: A Perspective on Evolution.* Translated by P. Woolley. Oxford: Oxford University Press, 1992.

Everett, Hugh. " 'Relative State' Formulation of Quantum Mechanics." *Reviews of Modern Physics* 29 (1957): 454–62.

Feynman, Richard. *"Surely You're Joking, Mr. Feynman!"* New York: Norton, 1985.

Fisher, Ronald A. *Statistical Methods and Statistical Inference.* Edinburgh: Oliver and Boyd, 1956.

Fitelson, Branden, Christopher Stephens and Elliott Sober. "How Not to Detect Design—Critical Notice: William A. Dembski, *The Design Inference.*" *Philosophy of Science* 66 (1999): 472–88.

Forrest, Barbara, and Paul Gross. *Evolution and the Wedge of Intelligent Design: The Trojan Horse Strategy.* Oxford: Oxford University Press, 2003.

Gerson, L. P. *God and Greek Philosophy: Studies in the Early History of Natural Theology.* London: Routledge, 1990.

Giberson, Karl, and Donald Yerxa. *Species of Origins: America's Search for a Creation Story.* Lanham, Md.: Rowman and Littlefield, 2002.

Goodwin, Brian. *How the Leopard Changed Its Spots: The Evolution of Complexity.* New York: Scribner's, 1994.

Gould, Stephen Jay. *Ever Since Darwin: Reflections in Natural History.* New York: Norton, 1977.

―――――. *The Panda's Thumb: More Reflections in Natural History.* New York: Norton, 1980.

Griffin, David. *Religion and Scientific Naturalism: Overcoming the Conflicts.* Albany:

State University of New York Press, 2000.

Guth, Alan. *The Inflationary Universe: The Quest for a New Theory of Cosmic Origins.* Reading, Mass.: Addison-Wesley, 1997.

Hacking, Ian. *Logic of Statistical Inference.* Cambridge: Cambridge University Press, 1965.

Harold, Franklin. *The Way of the Cell: Molecules, Organisms and the Order of Life.* Oxford: Oxford University Press, 2001.

Hartshorne, Charles. *Omnipotence and Other Theological Mistakes.* Albany: State University of New York Press, 1984.

Howson, Colin, and Peter Urbach. *Scientific Reasoning: The Bayesian Approach.* 2nd ed. LaSalle, Ill.: Open Court, 1993.

Hume, David. *Dialogues Concerning Natural Religion.* 1779. Reprint, Buffalo, N.Y.: Prometheus Books, 1989.

Hunter, Cornelius. *Darwin's God: Evolution and the Problem of Evil.* Grand Rapids, Mich.: Brazos Press, 2001.

Johnson, Phillip E. *Darwin on Trial.* 2nd ed. Downers Grove, Ill.: InterVarsity Press, 1993.

————. *The Wedge of Truth: Splitting the Foundations of Naturalism.* Downers Grove, Ill.: InterVarsity Press, 2000.

Kant, Immanuel. *Critique of Pure Reason.* Translated by N. K. Smith. New York: Macmillan, 1929.

Kauffman, Stuart. *At Home in the Universe: The Search for the Laws of Self-Organization and Complexity.* New York: Oxford University Press, 1995.

————. *Investigations.* New York: Oxford University Press, 2000.

Kolmogorov, Andrei. "Three Approaches to the Quantitative Definition of Information." *Problemy Peredachi Informatsii* (in translation) 1, no. 1 (1965): 3–11.

Kuhn, Thomas. *The Structure of Scientific Revolutions.* 2nd ed. Chicago: University of Chicago Press, 1970.

Küppers, Bernd-Olaf. *Information and the Origin of Life.* Cambridge, Mass.: MIT Press, 1990.

Kurzweil, Ray. *The Age of Spiritual Machines: When Computers Exceed Human Intelligence.* New York: Viking, 1999.

Lasker, Emanuel. *Lasker's Manual of Chess.* New York: Dover, 1960.

Leslie, John. *Universes.* London: Routledge, 1989.

Lewis, David. *On the Plurality of Worlds.* Oxford: Basil Blackwell, 1986.

Livingstone, David. *Darwin's Forgotten Defenders: The Encounter Between Evangelical Theology and Evolutionary Thought.* Grand Rapids, Mich.: Eerdmans, 1987.

Lloyd, Seth. "Computational Capacity of the Universe." *Physical Review Letters* 88, no. 23 (2002): 237901 (1–4).

Manson, Neil A. *God and Design: The Teleological Argument and Modern Science.* London: Routledge, 2003.

Margulis, Lynn, and Dorion Sagan. *Acquiring Genomes: A Theory of the Origins of*

Species. New York: BasicBooks, 2002.

McKeon, Richard, ed. *The Basic Works of Aristotle*. New York: Random House, 1941.

Medawar, Peter. *The Limits of Science*. New York: Harper & Row, 1984.

Meyer, Stephen C. "DNA by Design: An Inference to the Best Explanation for the Origin of Biological Information." *Rhetoric & Public Affairs* 1, no. 4 (1998): 519–56.

———. "DNA and the Origin of Life: Information, Specification, and Explanation." In *Darwinism, Design and Public Education*, ed. J. A. Campbell and S. C. Meyer. Lansing: Michigan State University Press, 2003.

Miller, Kenneth R. *Finding Darwin's God: A Scientist's Search for Common Ground Between God and Evolution*. New York: HarperCollins, 1999.

Mitchell, Melanie. *An Introduction to Genetic Algorithms*. Cambridge, Mass.: MIT Press, 1996.

Monod, Jacques. *Chance and Necessity*. New York: Vintage, 1972.

Nelson, Paul. "The Role of Theology in Current Evolutionary Reasoning." *Biology and Philosophy* 11 (1996): 493–517.

Newman, Robert, and John Wiester. *What's Darwin Got to Do with It? A Friendly Conversation About Evolution*. Downers Grove, Ill.: InterVarsity Press, 2000.

Nguyen, L., I. T. Paulsen, J. Tchieu, C. J. Hueck and M. H. Saier Jr. "Phylogenetic Analyses of the Constituents of Type III Protein Secretion Systems." *Journal of Molecular Microbiology and Biotechnology* 2, no. 2 (2000):125–44.

Numbers, Ronald. *The Creationists: The Evolution of Scientific Creationism*. New York: Knopf, 1992.

Olding, Alan. "Maker of Heaven and Microbiology." *Quadrant* 44, no. 363 (2000): 62–68.

Orgel, Leslie. *The Origins of Life*. New York: Wiley, 1973.

Orr, H. Allen. "The Return of Intelligent Design" (review of *No Free Lunch*). *Boston Review*, Summer 2002, pp. 53–56.

Paley, William. *Natural Theology: Or Evidences of the Existence and Attributes of the Deity Collected from the Appearances of Nature*. 1802. Reprint, Boston: Gould and Lincoln, 1852.

Pegis, Anton, ed. *Introduction to St. Thomas Aquinas*. New York: Modern Library, 1948.

Pennock, Robert. *Tower of Babel* (Cambridge, Mass.: MIT Press, 1999).

Pennock, Robert, ed. *Intelligent Design Creationism and Its Critics: Philosophical, Theological, and Scientific Perspectives*. Cambridge, Mass.: MIT Press, 2001.

Petroski, Henry. *Invention by Design: How Engineers Get from Thought to Thing*. Cambridge, Mass.: Harvard University Press, 1996.

Pierce, John R. *An Introduction to Information Theory: Symbols, Signals and Noise*, 2nd rev. ed. New York: Dover, 1980.

Polanyi, Michael. "Life Transcending Physics and Chemistry." *Chemical and Engineering News* 45 (August 1967): 54–66.

———. "Life's Irreducible Structure." *Science* 113 (1968): 1308–12.

Popper, Karl. *Conjectures and Refutations: The Growth of Scientific Knowledge.* New York: Harper & Row, 1965.

Quine, Willard. "Naturalism; or, Living Within One's Means." *Dialectica* 49 (1995): 251–62.

Ratzsch, Del. *Nature Design and Science: The Status of Design in Natural Science.* Albany: State University of New York Press, 2001.

Rea, Michael. *World Without Design: The Ontological Consequences of Naturalism.* Oxford: Oxford University Press, 2002.

Reid, Thomas. *Lectures on Natural Theology.* Edited by E. Duncan and W. R. Eakin. 1780. Reprint, Washington, D.C.: University Press of America, 1981.

Robertson, Douglas S. "Algorithmic Information Theory, Free Will and the Turing Test." *Complexity* 4, no. 3 (1999): 25–34.

Rolston, Holmes. *Genes, Genesis, and God: Values and Their Origins in Natural and Human History.* Cambridge: Cambridge University Press, 1999.

Royall, Richard M. *Statistical Evidence: A Likelihood Paradigm.* London: Chapman & Hall, 1997.

Ruse, Michael. *Can a Darwinian Be a Christian? The Relationship Between Science and Religion.* Cambridge: Cambridge University Press, 2000.

———. *Darwin and Design: Does Evolution Have a Purpose?* Cambridge, Mass.: Harvard University Press, 2003.

Sagan, Carl, *Contact.* New York: Simon & Schuster, 1985.

Sandbach, F. H. *The Stoics.* 2nd ed. Indianapolis: Hackett, 1989.

Savransky, Semyon. *Engineering of Creativity: Introduction to TRIZ Methodology of Inventive Problem Solving.* Boca Raton, Fla.: CRC Press, 2000.

Schaefer, Henry F. *Science and Christianity: Conflict or Coherence?* Watkinsville, Ga.: Apollos Trust, 2003.

Schleiermacher, Friedrich. *The Christian Faith.* Edited by H. R. Mackintosh and J. S. Stewart. Edinburgh: T & T Clark, 1989.

Searle, John R. *The Construction of Social Reality.* New York: Free Press, 1995.

Shannon, Claude, and Warren Weaver. *The Mathematical Theory of Communication.* Urbana: University of Illinois Press, 1949.

Shapiro, James. "In the Details . . . What?" (review of Michael Behe's *Darwin's Black Box*). *National Review,* September 16, 1996, pp. 62–65.

Shermer, Michael. *How We Believe: Science, Skepticism, and the Search for God.* 2nd ed. New York: Owl Books, 2003.

———. *Why People Believe Weird Things: Pseudoscience, Superstition, and Other Confusions of Our Time.* 2nd ed. New York: Owl Books, 2002.

Sherrard, Philip. *Human Image: World Image.* Ipswich, U.K.: Golgonooza Press, 1992.

Smith, John Maynard, and Eörs Szathmáry. *The Origins of Life: From the Birth of Life to the Origin of Language.* Oxford: Oxford University Press, 2000.

Smolin, Lee. *The Life of the Cosmos.* Oxford: Oxford University Press, 1997.

Sober, Elliott. "Testability." *Proceedings and Addresses of the American Philosophical Association* 73, no. 2 (1999): 47–76.

———. *Philosophy of Biology.* 2nd ed. Boulder, Colo.: Westview, 2000.

Spinoza, Baruch. *Tractatus Theologico-Politicus.* Translated by S. Shirley. Introduction by B. S. Gregory. Leiden: Brill, 1989.

Thaxton, Charles, Walter Bradley and Roger Olsen. *The Mystery of Life's Origin.* New York: Philosophical Library, 1984.

Van Till, Howard. "The Fully Gifted Creation." In *Three Views on Creation and Evolution,* ed. J. P. Moreland and John Mark Reynolds, pp. 161–218. Grand Rapids, Mich.: Zondervan, 1999.

———. "Science and Christian Theology as Partners in Theorizing." In *Science and Christianity: Four Views,* ed. R. F. Carlson, pp. 195–234. Downers Grove, Ill.: InterVarsity Press, 2000.

Weatherford, Roy. *The Implications of Determinism.* London: Routledge, 1991.

Wells, Jonathan. *Icons of Evolution.* Washington, D.C.: Regnery, 2000.

Whitehead, Alfred North. *Process and Reality.* New York: Free Press, 1978.

Wiener, Norbert. *Cybernetics.* 2nd ed. Cambridge, Mass.: MIT Press, 1961.

Wilkins, John, and Wesley Elsberry. "The Advantages of Theft over Toil: The Design Inference and Arguing from Ignorance." *Biology and Philosophy* 16 (2001): 711–24.

Witham, Larry. *By Design: Science and the Search for God.* San Francisco: Encounter Books, 2003.

Wittgenstein, Ludwig. *Culture and Value.* Edited by G. H. von Wright. Translated by P. Winch. Chicago: University of Chicago Press, 1980.

Wolpert, David H., and William G. Macready. "No Free Lunch Theorems for Optimization." *IEEE Transactions on Evolutionary Computation* 1, no. 1 (1997): 67–82.

Woodward, Thomas. *Doubts About Darwin: A History of Intelligent Design.* Grand Rapids, Mich.: Baker, 2003.

Wright, Ernest Vincent. *Gadsby.* Los Angeles: Wetzel, 1939.

Yanaras, Christos. *Elements of Faith: An Introduction to Orthodox Theology.* Edinburgh: T & T Clark, 1991.

Yockey, Hubert. *Information Theory and Molecular Biology.* Cambridge: Cambridge University Press, 1992.

Ziman, John, ed. *Technological Innovation as an Evolutionary Process.* Cambridge: Cambridge University Press, 2000.

Index